T0073548

How the Mind Changed

Also by Joseph Jebelli

In Pursuit of Memory: The Fight Against Alzheimer's

How the Mind Changed

A HUMAN HISTORY *of* OUR EVOLVING BRAIN

JOSEPH JEBELLI

Little, Brown Spark
New York Boston London

Little, Brown Spark
Hachette Book Group
1290 Avenue of the Americas, New York, NY 10104
littlebrownspark.com

First North American Edition: July 2022
Originally published in the United Kingdom by John Murray, July 2022

Little Brown Spark is an imprint of Little, Brown and Company, a division of Hachette Book Group, Inc. The Little, Brown Spark name and logo are trademarks of Hachette Book Group, Inc.

ISBN 9780316424981
An LCCN for this book is available from the Library of Congress.

Printing 1, 2022

LSC-C

Printed in the United States of America

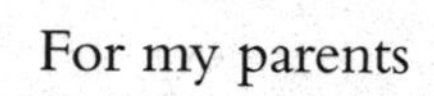
For my parents

Contents

The Brain—is wider than the Sky—
For—put them side by side—
The one the other will contain
With ease—and You—beside—

Emily Dickinson, *c.*1862

Introduction: The Pearl Inside the Oyster

Our minds are changing. They have been changing for nearly 7 million years. What these changes were, how they affect us today, and where they may lead us in the future is the subject of this book.

When we think of minds changing we usually think of psychological changes that affect our moods and outlook. Or neurological changes following a head injury or during an illness. But the changes I am interested in run much deeper. They span the evolutionary history of our early human ancestors and shape every aspect of who we are – our emotions, our memories, our languages, our intelligence and, indeed, the very fabric of our cultures and societies. It might not feel like it, but we are all heirs to millions of years of brain evolution: countless trial-and-error experiments in our mind's relationship with the natural world. As a result, we are cleverer and more interconnected than our forebears ever imagined.

We all intuitively understand that the brain is the most important organ in the body. We use it to think, feel, decide, act, move, and control our breathing and heart rate. We could lose our limbs and other organs yet think and feel as if nothing had happened. But if we so much as lost a piece of brain tissue the size of a grain of sand, our personality and behaviour could alter dramatically. Like fingerprints, brains are unique in each individual. The brain is a dancer and a poet, a teacher and a human rights campaigner. The brain is how we construct reality or engage in fantasy. The brain defines us as humans and distinguishes us from other animals. It is also what we most desire to enhance with future technology.

For as long as I can remember I've been fascinated by the brain. When I was a child, I routinely hassled my parents for the newsagent's shiny new copy of *New Scientist*. Sometimes I got lucky, and would take the treasured item to a quiet spot in the attic, crawl into the little wooden nook beneath the skylight, flick the pages to the first neuroscience article I could find, and stare in wonderment at the secrets it revealed. I didn't really understand them, of course, but I was transfixed by their endless riddles, their striking images, their unapologetic ambition.

The obsession crystallised in the final year of my undergraduate degree. There was a moment in one of my first lectures on 'action potentials' – electrical impulses in the brain – when I asked my professor, 'So everything that I am, all my thoughts and feelings, all the things I see and hear and touch, they're all just neurons firing these action potentials?' 'Yes,' he said. 'It's incredible, but yes.' I was dumbstruck. It was the first time I connected my experience of the world with the three pounds of matter floating inside my head. Four years later, I started publishing neuroscience articles of my own, tinkering with neurons and synapses in the laboratory, finally satisfying the cravings of my brain-obsessed youth.

As a neuroscientist, I think about the brain pretty much all the time. For more than a decade I have tried to discover, within my specific area of research, a small part of the puzzle of how individual neurons function – how the 85 billion cells in my brain allow me to perceive pleasure and pain, or remember the thousands of faces from my past. But throughout my research, a basic question has lingered at the forefront of my mind. *Why* have we ended up with the brains we have?

A hallowed list of explorers, philosophers, scientists and writers including Charles Darwin, Alfred Russel Wallace, Stephen Jay Gould and Jared Diamond have attempted to capture the wonder of human evolution. The humanist Julian Huxley wrote that evolution is 'the most powerful and the most comprehensive idea that has ever arisen on earth.' Nietzsche, Russell, Popper and Chomsky have all pondered the theory's wider significance. 'Nothing in biology makes sense,' said the Russian geneticist Theodosius Dobzhansky, 'except in the light of evolution.'

When Richard Dawkins called evolution 'the greatest show on earth', he was almost right. Hidden within, like the pearl inside an oyster, is something far more extraordinary: an evolving, changing mind. No other life form on the planet has generated a brain like ours – a brain spawning, to take but a handful of examples, ancient Greek philosophy and the music of Queen, Caravaggio's art, 7,000 living languages, and the cultural and scientific accomplishments of the Persian Empire. To study the brain is to study the essence of what makes us human. The evolution of the human brain is unique in that it makes substantial leaps forward. It's now thought this is because the brain relies on social changes that are adaptive and selective – a process scholars call cultural evolution. In this way, the human brain is an on-going project and the only brain on earth shaping its own evolution.

Understanding the history of the human brain helps us better understand our brains today. It allows us to make sense of our brains' daily functions, to explain why we love, feel joy, daydream, sleep, and think about thinking. This evolution isn't all positive, of course. By connecting our past with our present we also learn why we fight, feel jealous, deceive and hate. But even that understanding is important. Today, with all our digital technology and earthly distractions, we often lose sight of why we think the things we think. I want to change that. I believe that by understanding our long family tree of brains, we can change our habits and make better decisions in life. Such an understanding can provide us with important insights into human behaviour and modern society, allowing us to put aside our differences and come together to celebrate our shared evolutionary past.

There are many ways of telling the brain's story, with scholars from almost all scientific disciplines weighing in. There are accounts of the brain from anatomists, anthropologists, archaeologists, geneticists, genomicists, primatologists, psychologists, zoologists and, most intimately, from individuals who share their stories of the way their brain affects their lives. The history you hold in your hands will touch on all these fields, but like all histories is selective in parts. Natural histories are at the mercy of scientific breakthroughs, and the speed of these breakthroughs is accelerating more than at any time in the history of science. The more we investigate – and this is what I love about science – the more we uncover what we don't know, and this spurs us on to ever greater discoveries.

Our brains are the only brains we know, of course. We may never know what it is like to be a bat or if whales, with their colossal brains, view us with the same ceaseless wonder as we do them. But because all life and all brains evolved from common ancestors, they share some of the same characteristics. Thus our history is also the shared history of all life on the planet. And in a time of human-induced climate change and ecological disaster, it is incumbent on us to understand, acknowledge and respect this shared history in order to find our rightful place in nature. There are no more untouched places left on earth; if we want to evolve better minds, we must look within ourselves.

To tell this story, the book is divided into three broad themes. The first is an examination of the brains of the earliest humans, our most primal hardwiring, and the sequence of neurological and historical events that led to the human brain. Here, I will explore the most amazing brain capabilities that we take for granted: our creation of feelings, our sense of togetherness, and our capacity for remembrance. For the second theme of the book, I dive into the realms of higher cognition, exploring the origin of intelligence, language and consciousness. I reveal the extraordinary untapped powers of our brains, and how with societal change we can unlock them. The third and last theme of the book is the future of human brains and attempts to answer questions such as: will autism and neurodiversity change human life? Can humans free their minds from the confines of biology and achieve digital immortality? What are the implications of such advances, and will they solve the world's problems?

Above all, I want this book to deliver neuroscience's most uplifting message: that we all have the power to change our minds for the better. Because the truth is, none of us is born unintelligent, unfeeling, or unmoored from the world. These states are imposed on us by the societies we create. Minds make societies, but societies also make minds.

Barack Obama once declared the twenty-first century the golden age of brain science. I like to think of this book as the human tale that proves he was right. Think of it as a window into our brains' ancestral past, showing us the astonishing work that went into building such a unique brain, and how and why we came to look through the glass, darkly. There are amazing facts to be learned about the brain, facts that transcend our differences and change our behaviour for the better. For most of human history those facts were utterly alien to us – but now, thrillingly, we are finally learning where our brains come from, why they are so extraordinary, and how they are evolving today.

1

Building the Human Brain

Toumaï opens her eyes. Rays of sunlight break through the forest canopy. Lush green leaves cocoon the branches of her nest. She sees her family by her side and takes comfort in their well-being. Toumaï thinks and feels in whispers, for her mind is simple and ancient. She knows not what she is – but senses, perhaps, that she and her offspring are part of something much bigger than themselves. And indeed they are. For in the forests and woodlands of the Sahara, before natural climate change turned it into a desert, lived the earliest known human ancestor: *Sahelanthropus tchadensis* – otherwise known as Toumaï, an ancestor who lived 7 million years ago, 230,000 generations before you.

Toumaï had a tiny brain (350 cm^3), about the size of a child's fist. It's hard to say exactly what it looked like. A fossilised cranium, called an endocast, leaves only an impression of the brain on the skull; and comparing modern human brains to those of our closest ape relatives takes us only so far. Nonetheless, it was a brain that had to deal with many challenges. Sabre-toothed cats prowled the land; crocodiles patrolled the waters. Being in the middle of the food chain meant that Toumaï was always searching for food – and nearly always relying on scraps left by other, more lethal predators. With distinctive features including thick fur, strong arms, a sloping face and a prominent brow bone, Toumaï looked more ape than human. Whether she walked upright on two legs, a hallmark of human evolution, is unknown.

We first learned of Toumaï's existence on the morning of 23 March 2001, in the Djurab desert of northern Chad, west of the

Great Rift Valley. Ahounta Djimdoumalbaye, a Chadian student working with a group of French scientists, unearthed what we would later learn was a 7-million-year-old early-human skull. He named it Toumaï, which means 'hope of life', the name given to babies born in the Djurab before the dry season.[1] Paleoanthropologists had not witnessed a breakthrough of this magnitude since 1925, when the Australian anthropologist Raymond Dart uncovered a 3-million-year-old child's skull in Tuang, South Africa.[2] Toumaï, however, was even more impressive because she was the last common ancestor we shared with chimpanzees – the first chapter in the story of human evolution.

Humans first evolved in central Africa about 7 million years ago. After the dinosaurs perished, and mammals thrived and diversified, primates flourished in the treetops, where advanced social behaviours led to an increasing demand for greater cognitive power. Over time, different human species including *Ardipithecus*, *Australopithecus*, *Homo habilis*, *Homo heidelbergenis*, *Homo neanderthalensis*, *Homo naledi* and *Homo floresiensis* (to name a few) branched off from other apes and evolved with brains with unique characteristics. Some of these human species lived contemporaneously with one another; we *Homo sapiens*, for example, lived for a time alongside at least one other member of our genus, *Homo neanderthalensis*. Although you might feel superior to these now-extinct humans, it is important to remember that evolution has no aim; there is no inevitable march of progress. The famous cartoon of 'monkey to man' is the most misleading drawing in the history of science. It flies in the face of how evolution actually works – random mutations leading to non-random change – and is almost religious in its sentiment, imbuing us with what C. S. Lewis aptly called the 'snobbery of chronology'.

As a neuroscientist, I am familiar with human brains – which are a bit like large, squishy grapefruits – and I'll never forget the first time I held one.

It was September 2009, the start of term, and I was standing in a laboratory at University College London for an intimate lesson on brain anatomy. I looked at the brain in my hands. The light cast shadows over its wrinkled surface, and a pungent scent of preservative drifted up into my nostrils. It was heavier than expected, like a paperweight; it was beige with a pinkish tinge, like clay; it was soft but unyielding, like tofu. I slowly turned it over, eager to inspect it from every angle, before carefully passing it back to my professor. She pointed out the various lobes, cavities and ventricles, trailing her finger across all the regions we partly understand, plus all the ones we don't. This particular brain had belonged to a healthy elderly female – donated to help scientists understand how it works. Amid the bewitching folds of grey and white matter was a spellbinding tapestry of neurons and synapses, a cellular and molecular universe. Even today, the memory evokes a visceral sense of reverence and awe.

Alas I cannot – can never – hold a preserved version of Toumaï's brain. Neuroscientists are not detectives, but if we were, and if early human brains were the equivalent of missing persons, Toumaï's would be the ultimate cold case: a mystery so distant it lingers at the edge of comprehension, long unresolved yet tantalisingly open to new evidence. And such evidence is now emerging. Today paleoanthropologists have advanced software to create virtual imprints of ancient brains, which they use to show how and when particular brain shapes evolved. In January 2018 anthropologist Simon Neubauer and his colleagues showed that the brain started its journey as an elongated sphere (a bit like a small rugby ball), which gradually began bulging out into the globular shape of modern human brains.[3] This is a fascinating discovery. It suggests that the human brain existed as a kind of seedling version of itself, like a deflated beach ball waiting to be pumped with air. Eventually this expansion would give rise to our brain's four lobes: frontal, parietal, occipital and temporal, each housing different circuits for tasks such as thinking, speaking, seeing and feeling.

About 3.5 million years after Toumaï, another early human – *Australopithecus*, the 'Southern Ape' – lived in savannahs across Africa. Here, lions, leopards and hyenas posed the greatest threat. To survive, this ancestor needed to walk upright on two legs, freeing her hands to use primitive tools in the same way that chimpanzees use rocks to crack nuts and twigs to breach termite mounds. In 1974 scientists discovered a collection of fossilised bones in the Afar Triangle in Ethiopia belonging to a female member of this species, called *Australopithecus afarensis*.[4] The scientists liked listening to the Beatles' song 'Lucy in the Sky with Diamonds' while on the expedition, so they named her 'Lucy'.

If we met Lucy today, she would look more ape than human. She'd spend much of her time frolicking in the trees, enjoying a tree-dwelling lifestyle as well as a bipedal one. Crucial for this were her strong upper-limb bones, inherited leftovers from a close relative called *Australopithecus anamensis*, who was probably only a few hundred thousand years older. With a large body and extraordinarily long arms, Lucy would also have walked differently to modern humans, though what this movement looked like remains a mystery.

Although Lucy's brain was small (600 cm^3, about the size of a chimp's), it was starting to show subtle changes in shape and structure. By boosting the number of brain cells in a region called the neocortex, evolution gave Lucy her first glimpse of higher-order thinking, involving spatial reasoning, abstract thought and planning. The neocortex (Latin for 'new bark', because in the evolutionary tree of life 3 million years is practically brand new) is the folded outer layer of the brain, responsible for nearly all our higher faculties. It is unique to mammals, and is so important the astronomer Carl Sagan called it the place 'where matter is transformed into consciousness'. We can only guess what this was like for Lucy. I like to think of her thought processes as being similar to a young child's, not fully developed but impressive nonetheless.

What generated this increased brain size? As with so many evolutionary questions, genetics is at the root of the answer. In February

2015 a group of geneticists at the Max Planck Institute in Germany identified a stretch of DNA that appears to have triggered the boost in neocortex size.[5] The gene (rather unfortunately named *ARHGAP11B*) is highly active in human cortical stem cells (the progenitors of neocortex neurons), and, crucially, is not present in chimpanzees. It is uniquely human. Moreover, when the team inserted the gene into developing mice it made the animals' neocortex 12 per cent larger than usual. Their brains even started to display the characteristic folding pattern unique to the human neocortex. Toumaï and Lucy's brains might therefore have had similar folding patterns to our own.

Where this gene came from, we don't know. Evidence suggests that it appeared when a different gene partially copied itself – a process geneticists call gene duplication – after humans split from chimpanzees 7 million years ago. Of course, we still need to explain exactly what *ARHGAP11B* is doing to produce higher cognition in early humans. For now, that piece of the brain's evolutionary puzzle is conspicuously missing.

The Brain's Big Bang

Then came our genus (that is, a group incorporating multiple species), *Homo*, which emerged in East Africa about 2.5 million years ago – 1 million years after Lucy. For the brain, this represented a spectacular leap forward. Humans with brains of 900 cm^3 started to appear, followed by humans with a capacity of 1,000 cm^3. Soon afterwards, about 500,000 years ago, brain size ballooned in humans to a staggering 1,500 cm^3 – the size of a cantaloupe. Gone were the lifesaving attributes of other primates: thick fur, large muscles, a strong bite. Instead, evolution prioritised the brain. The neocortex expanded to occupy 80 per cent of the brain's mass, and new regions for intelligence, language, memory, creativity, self-awareness and conscious thought flourished. In an evolutionary heartbeat, a minuscule 0.014

per cent of the 3.5 billion years of life on Earth, the brain went from consuming 8 per cent of the body's energy to a massive 20 per cent, despite being a mere 2 per cent of total body weight. Though elephants and whales have bigger brains than humans, our brain is actually three times larger than what would be expected for an ape of our size. Relative to body mass, we have the largest brain of any living creature. If the human brain ever had a Big Bang, this was it.

You might wonder why this leap is so impressive, and not just another stage of development in the history of human brains. Though it may seem strange, our brain exhibited an unusual rate of change in evolutionary terms. It is not, as was once believed, just a linearly scaled-up primate brain. Our neurons are unique in many ways. They possess a unique genetic code, with at least thirty-two distinctly human genetic signatures shared across 132 brain regions.[6] They have unique membrane and synaptic properties, allowing them to boost their connectivity and computational power. Most important, they are more flexible and more malleable than the brains of our closest ancestors, giving them a competitive edge in cognition and learning ability. Acknowledging these facts is not a prescription for human exceptionalism; it is simply the recognition that with *Homo sapiens*, the brain changed dramatically.

Throughout *Homo* evolution, many types of human brain evolved simultaneously. One belonged to *Homo habilis*, which means 'handy man', in deference to their tool-making abilities. These humans were nomadic hunter-gatherers, living in small bands in the grassy plains of northern Tanzania. Another type belonged to *Homo erectus*, 'upright man', who thrived in Africa and Eastern Asia for so long (2 million years), they are what biologists call a chronospecies: a species that changes and improves without ever becoming a new species altogether – a biological time traveller. Quite how this human changed yet did not change enough to become a new species is an on-going riddle in anthropology, and so some scientists distinguish the African variant (*Homo ergaster*) from the Asian variant (*Homo erectus sensu stricto*). Several other brains also flourished at this point, including that of *Homo heidelbergenis*, *Homo neanderthalensis*, *Homo naledi* and *Homo floresiensis*, though we know very little about these species, what their brains looked like or how they lived.

Such variety arose because the brain, like any organ, is subject to the pressures of natural selection. The planet has withstood immense changes over the past 2.5 million years. The grasslands of Africa and Arabia, once shaded by thick woods and watered by torrential monsoon rains, morphed into a fierce desert. The Earth's unsteady orbit, shifting every 20,000 years, triggered an ice age every 100,000 years. Over a small period of geological time, the planet's surface warmed and cooled, warmed and cooled. And each period brought special problems that only specialised parts of specialised brains could solve. *Homo habilis* and *Homo erectus*, for example, probably possessed neural circuitry for advanced social cognition, allowing them to persist through the ice age by hunting in groups. But as the ice melted and groups swelled, species such as *Homo naledi* evolved circuitry for goal memory – the power to remember certain objectives and targets of action; in the ruthless environment of primate society, remembering who your allies were was critical. And on it went. Time waxed and epochs waned; random mutations fuelled non-random advantages (or disadvantages); and natural selection filtered out what worked for the brain and what didn't.

The human genome has experienced massive selection in the past few millennia – thought to be as significant as the artificial selection seen in domestic dogs, all of which trace their roots to a single group of grey wolves. Perhaps the most distinctive evolutionary change for our species was the loss of human body hair, with one theory suggesting that we went through a semi-aquatic phase (hence our slightly webbed hands), and another that our ancestors needed to keep cool when they migrated across the hot

African savannah. But body hair is one thing, the human mind quite another. So how was the change achieved? Recent studies suggest that at least a dozen new genes contributed to humans' advanced cognitive capacity, each one coding for a slightly different molecular spring, cog, gear and dial in the ever-expanding clockwork of the mind.[7] While this is clear evidence for the role of genetics in supersizing the human brain, the real causes can be found in the interplay between DNA and the environment. Two diverse theories, as interesting as they are surprising, go some way towards explaining what really happened.

In 2004, Hansell Stedman, a molecular geneticist at the University of Pennsylvania in Philadelphia, studied the genomes of people from across the world – including natives of Africa, Europe, Russia, Iceland, South America and Japan – and compared them to several non-human primates alive today, including gorillas and chimpanzees. While the results remain the subject of debate, it appears that a rare gene mutation severely shrunk the jaw and weakened the bite of our early human ancestors.[8] Biting and chewing are controlled by powerful muscles in the jaws of most primates, and genetic research shows that a gene called *myosin heavy chain 16* (*MYH16*) plays a crucial role in this kind of muscle contraction.[9] When chimpanzees bite, *MYH16* switches on, and the muscles apply a strong force over the skull, restricting its growth. When mutated, however, *MYH16* causes the jaw muscles to be eight times smaller than those of other apes. And this is now thought to be one of the reasons our species, *Homo sapiens*, has the largest brain of any primate. By eliminating the restraints of a bulky jaw, the human skull expanded, freeing the brain to grow to its modern size – three times the size of the average chimpanzee's.

'The first thing to note about the human brain is its size,' Stedman told me during our long conversation. 'I'm not suggesting that this mutation alone buys you a human brain, but it certainly could have started the brain's evolutionary process.' Because the cranium and jaw are both made of spongy bone, Stedman thinks the process was inevitable. 'Muscle sculpts bone,' he said. 'And the human skull has always been modified by the forces acting on it, like a river slowly changing the landscape over time.' He added that powerful jaws cannot coexist with powerful brains, due to the anatomy of cranial structures in primates. Consider the chimpanzee: its skull has no forehead; it has a large brow bone and a distinctly projecting face. This means its brain is forced into a small case like a tiger in a cage.

Stedman emailed me an image of a human and chimpanzee skull. The contrast between skull size and jaw size is stunning. It looks to be by design, but of course it isn't. When it comes to evolution, bone is just as haphazardly constructed as everything else. A good example is the spine. It evolved to be stiff for climbing and moving in trees. Then we walked upright and it curved inward to deal with the weight of the head. But all that extra pressure causes back problems and vertebral fractures and a whole host of other issues. Then there's the foot. No one would design it to have twenty-six bones. It's built that way because our ancestors needed flexible feet to grasp branches – but again, the pressure of walking upright causes problems such as ankle sprains, shin splints and Achilles tendonitis. The *MYH16* mutation in the jaw seems to be one of the few evolutionary instances where we got lucky. Really lucky. 'All the other primates' brains were constrained by their jaw muscles,' says Stedman. 'But *Homo sapiens* had nothing holding them back. They got smarter and smarter and smarter.'

It's easy to assume that something *within* the brain must have triggered its evolutionary ascendency. The idea that it all started with a genetic accident in the jaw is almost too surprising. But the truth – as we will discover throughout this book – is that evolution is a game of chance, not strategy. DNA mutates by various means – from sloppy cell division to climate change to interstellar cosmic rays. That's not to say that evolution is a bad engineer. Quite the opposite. The blind nature of the process, unshackled by the imaginative limits of human engineers, permits untold brilliance – the kind that allowed the pectoral fins of fishes to become the forelimbs of horses, the flippers of whales, the wings of birds and the arms of humans.[10] And a single mutation is all it takes: if a mutant organism produces just 1 per cent more offspring than its non-mutant rivals, it leaps from representing 0.1 per cent of the population to 99.9 per cent in just 4,000 generations, a mere 100,000 years. (To put that in context, if the height of the Empire State Building represented the history of the planet, 100,000 years would be a postage stamp at the top.) We don't know what caused the *MYH16* mutation – and may never know – but what's clear is that the revolving door between fate and happenstance is far more significant than has been previously recognised.

Another possibility is the invention of cooking, a unique human skill that saves energy by allowing us to expend less on chewing and digesting food. Cooking breaks down the connective tissue in animal flesh and dismantles the carbohydrates in plants, both of which improve absorption in the gut. Around 2.7 million years ago, before microwave meals and the dawn of the Big Mac, humans probably only cooked things such as wheat, root vegetables, fibrous fruit and, most importantly, meat: a complete protein containing all twenty amino acids and packed with energy-rich fat. In 2007 the American physiologist Stephen Secor showed that Burmese pythons fed a meal of cooked versus raw beef expend 23.4 per cent less energy digesting the food if the beef is cooked.[11] Others have found similar results across the animal kingdom.[12] Meat is also a good source of niacin (vitamin B3), a nutrient known to enhance brain development, and cooking increases niacin extraction rates from meat by 50 per cent.[13]

As humans ate more and more meat, they consumed enough calories to allow their gastrointestinal organs to shrink, which diverted even more energy to the brain. Essentially, humans swapped guts for brains. This theory, called the Expensive Tissue Hypothesis, is compelling. After all, the human gastrointestinal tract is only 60 per cent of the size it would be for an ape of our stature, and tiny guts have been observed in six other species of brainy primates, including capuchin and howler monkeys. One might think the gut was losing out in the deal, but as primatologist Richard Wrangham points out, cooking allowed humans to build an external stomach in which fire, rather than digestive enzymes, was used to break down food before it was processed by the body. 'It makes sense that we like foods that have been softened by cooking,' Wrangham writes,

> just as we like them chopped up in a blender, ground in a mill, or pounded in a mortar. The unnaturally, atypically soft foods that compose the human diet have given our species an energetic edge, sparing us much of the hard work of digestion.[14]

Evolutionary trade-offs are not unique to humans. Male howler monkeys, for example, have traded a loud roar for bigger testicles: the louder the roar in these primates, the smaller the testicles. Believe it or not, this is thought to help howlers mate. A tiny-balled male might struggle to attract a female, but having a ferocious roar will certainly help scare away any male competition. On the other hand, a big-balled male is so desirable he can afford to live in a group with other big-balled males, keep quiet and simply wait for the females to mate with him and his friends – which they do. Research tells us that trading guts for brains works because humans embraced a particular life history. Unlike other animals, humans extended their juvenile years, delaying reproduction until much later in their lifecycle. In consequence, humans had more time to shrink their guts and grow larger, more sophisticated brains before adulthood. The postponement of development – called neoteny in evolutionary biology – is especially active in the human brain. A recent genetic analysis found that 40 per cent of genes linked to the development of the prefrontal cortex only become active well into adolescence.[15]

If cooking and our resulting micro guts really did provide the energy needed for our macro brains, we would expect to find evidence dating back at least 300,000 years, when *Homo sapiens* emerged, of the most vital ingredient – fire. The evolutionary advantage fire gave humans was immense. In addition to providing heat to cook food, fire brought light, warmth, protection from deadly predators and a meeting point for groups to socialise, share stories and form meaningful bonds. There is evidence that humans controlled fire (scorched earth, charred bones, charcoal, ash, stone hearths) from 800,000 years ago, in Israel – so well before the brain's Big Bang 500,000 years ago. Some scholars point out that seared clay and burned stone tools have been found at campsites dating back 1.5 million years, at Olduvai Gorge in Tanzania and Koobi Fora in Kenya. Others posit that 1.4-million-year-old burned clay has been found in the Baringo Basin of Kenya, and that 1-million-year-old charred animal bones have emerged at Swartkrans, South Africa. But all such evidence comes with a big caveat: were these fires human-made or natural? After all, lightning and volcanoes are just as able to ignite a fire. But even if fire was tamed millions of years ago, we are still left with the mystery as to when humans prepared the first cooked meal.

One thing is certain: between 500,000 and 2 million years ago something extraordinary happened. The brain of *Homo erectus* alone was suddenly a colossal 25 per cent bigger than that of its predecessor, *Homo habilis*. On the scale of things, this is an even greater evolutionary change than the whale graduating from walking on land to a life in the water in only 10 million years.

The *MYH16* mutation and the advent of cooking are only two in what will probably turn out to be a long list of theories about why the human brain is so big. Indeed, as I write, a gene called *ZEB2* has been found to be an important molecular switch in brain development, nearly doubling the number of neurons in the human brain compared with that of other great apes.[16] In the years ahead, scientists will increasingly rely on such evidence to get a clearer picture of what occurred. Discovering how the brain evolved is the most ambitious excavation of its kind, because the mind, as George Elliot wrote, 'is not cut in marble – it is not something solid and unalterable. It is something living and changing.'[17]

An Alternative History

But what if it didn't change? What if fate stepped in and stopped the brain ascending the evolutionary ladder? I've contemplated this sort of question before. When I was a research scientist at the University of Washington, Seattle, my colleagues and I would attempt to understand something about the brain by imagining its absence. What if, for example, the fatty myelin sheaths that wrap around neurons disappeared? (Answer: the brain's electrical impulses slow down and the symptoms of disorders such as multiple sclerosis appear.) What if the protein molecule 'tau' suddenly vanished? (Answer: the neuron's internal skeleton collapses and the symptoms of Alzheimer's disease appear.) It became something of a game, with brownie points reserved for the scientist who posed the most intriguing question.

To understand how miraculous the invention of the human brain was, we can ask what might have happened if, 2.5 million years ago, there had been no *MYH16* mutation in the genus *Homo*, or that some variety of natural disaster – an earthquake, an asteroid, a plague – had wiped out every variety of *Homo* on the African subcontinent. No mutant, no massive brain, no momentous increase in cognitive power and cultural sophistication. The human race as we know it ceases to exist.

A good bet for a brainy primate that might have taken our place is the Indonesian species *Homo floresiensis*, a miniature human-like ape that rarely grew over three feet and had huge, fuzzy feet. In 2003, when scientists found remains of the tiny humanoid in a cave on the island of Flores, one of them exclaimed, 'Holy shit, hobbits!', which they have been nicknamed ever since.[18] It's thought we *Homo sapiens* wiped out the hobbits when we invaded their island some 50,000 years ago. No surprises there.

But now, with *Homo sapiens* gone, hobbits reign supreme. Their brains are small (about the size of Lucy's, *Australopithecus*), but they can still build stone tools, hunt elephants, use fire and fend off giant komodo dragons. So they're certainly intelligent enough to migrate out of Indonesia. Facing no real competition, they make their way by raft to Australia, Malaysia, Vietnam and the Indian Subcontinent, colonising as they go, seeding new hobbit societies and turning them into warring tribes that compete for land and resources. Within a few million years, during a period of intense migration and plummeting temperatures, hobbit colonies in the northern hemisphere spawn a new generation of weather-hardened 'woolly' hobbits, who huddle around fires on the ice sheets of Eurasia, surviving on seal meat and the occasional mammoth, pondering where next to call home.

But that's it. With no mutations boosting rapid brain evolution, the hobbits' brains remain unchanged. And they continue to live as nothing more than clever apes – wandering nomads, waiting to inherit a genetic recipe that never comes, oblivious to the scientific and civilisational marvels that nature had planned for their long-lost *Homo sapiens* cousins. No doubt this alternative world would have been an infinitely more primitive place, but it is one that could have easily existed.

When viewed in the light of this evolutionary game of chance, it is even more extraordinary that it is our species, *Homo sapiens*, 'Wise Man', whose brain triumphed above the rest, after it entered into existence some 400,000 years ago. The era into which *Homo sapiens* emerged was one of extreme ecological instability. African megadroughts depleted the land's fresh water; vanishing grasslands diminished the number of animals available to hunt. To survive and flourish, *Homo sapiens* spread across the world, encountering and interbreeding with other species of *Homo* along the way. One was the Neanderthals, early humans who split from modern humans about 500,000 years ago and went extinct around 40,000 years ago. In recent years, we have learnt that modern humans living outside Africa carry around 2 per cent Neanderthal-derived DNA, the result of Neanderthal–human interbreeding 40,000 to 60,000 years ago. Another was the Denisovans, cousins of the Neanderthals who thrived across Asia between 500,000 to 30,000 years ago. Humans living today in Oceania, particularly Papua New Guinea and Australia, possess up to 6 per cent Denisovan DNA from ancient interbreeding. It's thought that Southeast Asians once interbred with *Homo erectus* and possibly *Homo floresiensis*, the 'hobbits'. Species such as *Homo naledi* and *Homo rhodesiensis* also probably coexisted with what became modern humans (though whether they interbred is unknown). When it comes to the brain, this genetic mixing between different humans may have contributed to its astonishing size and complexity.

Today our advanced social behaviours have led to an increasing demand for even greater cognitive power, sending the modern human brain into hyperdrive. While the underlying changes remain hidden from our conscious minds, we now have the tools – advanced microscopy and molecular genetics – to spot them in unprecedented detail. Just as we use the rings in a stump to learn the age of the tree (the more rings it has, the older it is), so too can we use the mosaic of neurons, the constellation of synapses and the tributaries of molecules to learn the age of the brain and the transformations it has seen. As we will see, understanding these changes is the key to improving our behaviour, our health, our environment, our education and our morality.

Homo sapiens' brain anatomy – our brain anatomy – is breathtaking in its complexity. Though it may look like a homogenous ball of grey and white matter, the brain has three basic parts. The largest part, the cerebrum (Latin for 'brain'), sits at the top. Whenever you see a picture of the brain, with its characteristic folds and wrinkles that make it look a bit like a walnut, you are looking at the cerebrum. It controls all our higher intellectual functions and helps knit together all the sensory information flooding into our minds from the outside world. It divides into two hemispheres, the left and the right, which for unknown reasons control opposite sides of the body. These hemispheres communicate with each other through a thick bundle of nerves called the corpus callosum (Latin for 'tough body'), which is larger in musicians, ambidextrous people and homosexual men.

The cerebrum can be divided again into four lobes: frontal, temporal, parietal and occipital. The frontal lobe is essentially the control panel of our personality: important for thinking, feeling, speaking, judging, planning, socialising and controlling our sexual behaviour. The temporal lobe is a different beast altogether; it's a kind of translator for auditory information, turning every signal received from the ear into a message that our brains can understand. The parietal lobe is all about sensation: our sense of touch, temperature and the location of our body in space. Damage to the parietal lobe can cause depersonalisation disorder (DPD), a condition where people feel completely detached from their own bodies, as though they are living their lives on autopilot. The occipital lobe is the seat of vision: it's here that everything we see – all the shapes, colours and movements – are analysed and interpreted to produce a seamless cinematic-like projection of the world. If even one small part of the occipital lobe is lost – to a stroke, say – a person can experience symptoms ranging from the inability to recognise faces to seeing the world in snapshots.

At the back of the head, where the spine meets the brain, is the cerebellum (Latin for 'little brain'), which houses more than half of the brain's neurons. For a long time it was thought that the cerebellum's only role was to control voluntary movement, the kind you need to pass a breathalyser test. But with so many neurons – the result of the cerebellum expanding over evolutionary time – it came as little surprise to learn that it also does a great deal of thinking. We just don't know what sort of thinking. Some believe it checks and corrects thoughts in the same way it checks and corrects movements;[19] others believe it acts as a kind of editor not only for thoughts but for emotion, language and memory.[20] As Nico Dosenbach, a neurologist at Washington University, puts it: 'We have an explanation for all the bad ideas people have when they're drunk. They're lacking cerebellar editing of their thoughts.'

Beneath the cerebrum and in front of the cerebellum is the brainstem, the gateway between the brain and the body. It oversees all the bodily functions that we're not normally aware of: sleeping, breathing, swallowing and controlling the heartbeat. It is the most ancient part of the brain, evolving more than 500 million years ago, and is crucial for keeping us alive. When snipers need an instant kill, they aim for the brainstem.

Dozens of smaller structures are spread throughout the brain, with names as exotic as the amygdala, hypothalamus, hippocampus, telencephalon and habenular commissure, to name a few. Each is specialised for a particular aspect of brain function and each, as we will see, has its own evolutionary history. Fanning out across all this are the brain's blood vessels. Though tiny in size they are amazingly long: if they were laid out in a line they would measure more than 60,000 miles, enough to circle the globe.

For all its staggering sophistication, the brain is a remarkably vulnerable organ. The skull fails to protect it against serious falls or collisions. A tiny clot, or stroke, can wipe out whole chunks of brain tissue. Viruses can infect it and trigger an immune response, both of which can be fatal. Tumours can burrow in and hollow it out. Psychiatric afflictions such as schizophrenia, depression and bipolar disorder can traumatise it. And neurodegenerative diseases such as Alzheimer's, Huntington's, Parkinson's and multiple sclerosis can bore holes in it until it resembles Swiss cheese. According to the World Health Organization (WHO), one in four people will be affected by a neurological illness at some point in their lives.

You might be wondering why the brain, unlike other tissues in the body, doesn't seem to heal itself. Santiago Ramon y Cajal (1852–1934), a Spanish physician and one of the founding fathers of neuroscience, was convinced that new neurons are only added to the brain before birth. Scientists have been investigating this and recent research shows that he was wrong. At least two brain regions can produce new neurons: a part of the hippocampus known as the subgranular zone (SGZ) and an area flanking the brain's many cavities, or ventricles, known as the subventricular zone (SVZ).[21] The problem is that the birth of new neurons in the mature brain, so-called neurogenesis, replaces very few of the neurons lost to common brain diseases such as Alzheimer's and stroke. For this reason, scientists are zealously pursuing a way to boost the brain's healing powers artificially.

Nevertheless, the brain has evolved with some remarkable tricks up its sleeve. In 2007 *The Lancet* reported the case of a forty-four-year-old French man who had been living normally despite the fact that he was missing 90 per cent of his brain.[22] A disease called hydrocephalus had been silently destroying his brain since childhood, leaving only a thin outer layer of brain tissue and a gaping hole in his head. A civil servant with a wife and two children, the man was leading a healthy life and had only gone to his doctor complaining of mild weakness in his left leg. His brain had literally reorganised itself, radically changing to compensate for what would otherwise have killed him or left him in a vegetative state. In recent years, scientists have given a name to this phenomenon: neuroplasticity.

This is the brain's ability to rewire its synaptic connections, reorganising itself in response to new circumstances or changes in the environment. Discovered by Michael Merzenich in the early 1970s (when he was in fact trying to prove the exact opposite: that the brain is fixed and, thus, unchangeable), neuroplasticity is thought to exist in some form in all primates, and is the driving force behind the brain's evolution – a turbo-charge for its ability to change. This remarkable capacity for adaptation is the reason we can learn a new language, play a new instrument and navigate a new world. It's the reason stroke victims can recover. It's our brain's perpetual second chance.

Today, scientists are desperately trying to unlock the secrets of neuroplasticity to treat all manner of brain diseases. For a long time, I was one of them. Back in the early days of my research, my interests lay in understanding the more fundamental properties of nervous systems, including their regenerative abilities. So every week, my lab mates and I would band together to produce a culture of neurons (taken from a rat) and watch how they behaved under different conditions. We had no grand hypothesis. We were just doing what all neuroscientists do at that stage of training – frolicking in the lab, learning by failing, trying not to break anything. It was a thrilling time and a brilliant opportunity to ask basic research questions. Something I wanted to know was: how did the neurons stand up against a small dose of the bacterial toxin lipopolysaccharide (LPS)? LPS decorates the surface of many bacteria, such as Salmonella and E. coli, and evolved to shield bacteria from harm as well as help them escape the host's immune system. (We had a lot of LPS in our lab because other scientists were using it to study the neuroinflammatory disorder multiple sclerosis – MS.) My idea was a simple experiment, and since relatively little is known about how neurons respond to LPS, I figured it would be a nice way to observe the resilience of brain cells in real time. I treated my cells, returned them to the incubator, then checked on them a day later.

The LPS had had a devastating effect. More than half the neurons were dead, their microscopic carcasses reduced to flotsam and jetsam in the culture medium, shrivelled husks of their former selves. Those that survived looked healthy enough, though I couldn't assess their internal state with just a microscope, and I assumed they'd be dead soon too. I remember feeling saddened by the result; clearly neurons weren't that resilient. How many of my own had I killed after a big night out, I wondered. How many was I harming *right now* just by worrying about their terrifying fragility?

The next day, I checked on my remaining cells and couldn't believe what I saw. Not only were they alive – they were thriving. Each of the surviving neurons was sprouting new projections, forming new synapses and making new connections. Like the forty-four-year-old French man missing 90 per cent of his brain, they had reorganised themselves in response to the injury, enhancing and seemingly ramping up their function to offset the loss of their neighbouring cells. It was a breathtaking moment, something I'd only ever read about until that day. From then on, neuroplasticity was an endless source of fascination to me. More than anything, the phenomenon is a striking reminder of how malleable our brains are, and how evolution has made even the most remarkable brain changes possible.

Such marvels of anatomy are not unique to *Homo sapiens.* Other species have brain anatomies that are beautiful, vulnerable and often surprising. The brain of a bird is completely smooth and uniform, with no wrinkles or lobes to speak of. The brain of a spider is so large that it fills their body and spills into their legs. Teleost fish, the most diverse of all fish, can continuously regenerate their brains in adulthood. Once a sea squirt finds a nice spot to call home, it eats its own brain. Such novelty exists for the same reason we see novelty across species: diverse environments promote a diversity of brains adapted to them.

Cultural Evolution

Among the brain's characteristics that might be deemed uniquely human is its ability to accumulate culture. We call this phenomenon cultural evolution, and it's a big deal. Unlike purely genetic evolution, in which traits are passed directly between parents and their offspring, cultural evolution allows any member of a population to inherit traits from any other member. It is still considered a form of Darwinian evolution; the key difference is that traits are acquired not by DNA, but by the relationships between human beings. We see cultural evolution happening all around us, all the time. When a teacher explains the fundamentals of physics to her students, she is passing on information that will help them flourish later in life. When society demands the equal treatment of women, a new selection pressure to copy and imitate such behaviour is transmitted culturally. And when a politician or activist calls for action on climate change, those who listen are inheriting knowledge crucial for the planet's – and consequently their own – survival.

But steady on. Isn't this all just dependent on the norms and social conventions of the day? What does the way society treats women or the climate possibly have to do with the cognitive advancement of our species? The answer is: everything. In fact, cultural evolution may be more important than biological evolution. To understand why, let's imagine two groups of people: the 'Savs' and the 'Dims'. Both are defined not by their DNA but by their respective cultures. The Savs promote an inclusive culture in which every member of their society is treated fairly and human rights are upheld. The Dims, on the other hand, enforce a culture of gross inequality, heavily curtailed freedoms and the persecution of minorities. Needless to say, the offspring of each group will inherit their group's unique cultural traits by way of teaching, imitation and other forms of cultural transmission. Thus, the Savs will continue to be cultural Savs, and the Dims cultural Dims.

Yet what if there is something about the Savs' pro-equality stance that makes them better at performing some task – creating a strong economy, say – than the Dims? We know that people born in wealthier countries live longer than those born in poor countries (a cursory glance at the Preston curve demonstrates this).[23] We know too that gender equality boosts economic growth and shrinks mortality rates.[24] In consequence, the Savs produce healthier offspring who are more likely to reproduce. They are the fitter group, and the Dims know it. Try as they might to convince their citizens otherwise, the Dims will lose many members of their society over time. Like the Afghans fleeing the Taliban and the Venezuelans fleeing the Maduro regime, they will vote with their feet and migrate to the Savs' more enlightened world.

This is how the brain empowers evolution by culture. By simply running the cognitive software of good ideas – in our example the idea of freedom over tyranny – the human mind became capable of improving the biological fitness of entire groups. DNA replication became almost an afterthought. With the mind of *Homo sapiens*, an entirely new selective pressure was born.

There is a sense of wonder and bewilderment when we consider the brain that evolution has built for us. You only have to look at the other great apes – the orangutans, gorillas and chimpanzees – to see the vast difference between their way of life and our own. Of all the primate brains to evolve, ours was not just the best innovator, the best problem solver, the best society-builder; it was also the fastest and the most energy efficient. In its exclusive identity and marvellous eccentricity, the human brain became our crowning achievement. We really do have a brain like no other. And so let us begin by understanding one of the oldest parts of the brain: the emotional brain.

2

Inventing Emotions

Four million years ago, the hairy, four-feet tall human ancestor known as *Australopithecus anamensis* wandered the woodland habitat of East Africa. She walked fully upright on two legs, but looked more ape than human. She was brawny and weighed nearly eight stone, but was more prey than predator.

Many environmental challenges surrounded her, hampering her abilities to gather plants and hunt animals, and procreate to keep her species alive. To make matters worse, her own evolutionary ancestors had a conspicuous record of being killed by stealthy leopards and other big cats. Defence against harm was critical for her to survive: her number one priority.

Fortunately, millions of years of evolution had equipped her body with sensory neurons to detect harm and motor neurons in order to trigger a response (ideally in her legs). Known as 'survival circuits', these sensory neurons are as old as life itself – even single-celled organisms such as bacteria sprout sensors to 'feel' around for harmful chemicals and can physically retract from a threat if necessary. The problem was that her response to danger was still rather unsophisticated. Survival circuits are useful at the moment of attack to help you run – or, if she was brave, put up a fight – but a leopard is a leopard.

What she really needed was an intelligent warning system, a way to scrutinise threats rigorously and react before it was too late. And that's exactly what evolution provided. First appearing in fish nearly 500 million years ago, the human brain formed a region called the amygdala, an almond-shaped structure buried deep in each temporal lobe. (For a memorable – albeit macabre – way to appreciate its location, picture an arrow entering your left ear and another entering your left eye: the point where the lines converge, low down and towards the outer edge of the brain, is the amygdala.) As with most brain structures, you have two amygdalas, one on each side of your brain.

Because evolution is an excellent recycler, always building on what is already available, the amygdala actually developed from our most primitive sense – the sense of smell. Sitting just beneath the amygdala is a region called the olfactory bulb, which controls smell. As our brain grew more complex, evolution simply layered the amygdala on top. This was a smart move by the brain: in positioning itself this way, it could be hooked up to receive signals from our other four senses: sight, sound, taste and touch.

One might think that processing so much information would be a slow, cumbersome task for the brain. But the pathway linking the amygdala to the senses is shorter than the pathway linking it to the neocortex, the home of higher-order thinking that evolved much later. This means that we don't have to think or reason our way through a situation that demands an instant emotional response. It's why you know a stranger following you down an alleyway doesn't feel right; why unexpected sounds in your house at night ring alarm bells in your head.

Medical literature is replete with examples of what can happen when the amygdala is jeopardised. One woman who had amygdala damage (due to an unusual genetic condition called Urbach-Wiethe disease) became completely immune to fear. Known only as SM to protect her identity, she was unfazed by exposure to snakes, spiders and horror films, and has been held at gunpoint and knifepoint and still felt no fear for her life.[1] Patients with amygdala damage report less fear of gambling and can develop something called affective blindness, the inability to identify the emotional expression of faces correctly.

For *Australopithecus anamensis*, our furry, modestly sized ancestor, this brain change meant they could finally learn to associate certain things in their environment – a rustle in the trees, the scampering of monkeys, the smell of a big cat – with imminent peril. And so from the paralysing depths of fear, our ancestors emerged defiant, reborn with emotional intelligence and a neurological edge to survive. As humans evolved to live in nomadic foraging bands, the amygdala adapted, generating five more primal emotions: anger, surprise, disgust, joy and sadness.

And when, millions of years later, *Australopithecus anamensis*' descendent, *Homo sapiens*, generated culture and civilisation, the amygdala adapted again, sparking sophisticated kinds of emotional processing such as forgiving the wrongs of those we love, or confronting and accepting loss. Exactly how it does this remains unclear, but it's thought that neurons in the amygdala modulate the activity of the neocortex, which in turn modulates our emotional behaviour. The relationship between the amygdala and the neocortex allowed our brains to experience new kinds of emotional complexity. Despite being more than 5 million years old, the amygdala thrives in the modern world. 'If all the world's a stage,' declared the cognitive scientist Adam Anderson, 'then the amygdala may be the emotional spotlight.'[2]

For Amanda Walker, a forty-eight-year-old interior designer from Oxford, that emotional spotlight shattered when a stroke tore through her amygdala. Now, she lacks the capacity to feel joy or love or sorrow. She has a condition called alexithymia, a kind of emotional colour-blindness. Asked how she feels about her predicament, she responds, 'I can't answer that. There *is* nothing to feel.'

In February 2021 I video-called Amanda at her home in Jericho, a laid-back suburb near central Oxford. She rarely leaves the house: her condition makes it difficult for her to interact with people, and she's grown tired of having to explain herself to the outside world. She lives alone, her two cats being all the company she wants, and spends her days devouring books and painting abstract art. Friends visit from time to time. A private carer checks up on her once a week.

'Thankfully I'm still able to do my job online,' she told me.

> My boss has been very supportive, very patient. It takes me much longer to design anything now. Interior design relies on emotion a lot more than people think. Everyone knows how a beautiful home affects your mood, but the process itself is just as emotional.

Amanda now thinks of life as a performance. She puts on a smile for friends' photos; she knows to laugh when it's meant to be a joke; she scrunches up her face when hearing bad news. 'It feels like I'm acting,' she admits.

> Like I'm faking being a person. Sometimes I get a flash of real feeling, as if there's *something* there that my brain just can't grab hold of. But most of the time I'm following a script. My brain is basically asking, 'What response is socially acceptable for this situation?' and then I just do that. I'm not suffering in the way someone with bipolar or schizophrenia is suffering, but it's taken a lot of enjoyment out of my life.

I wondered if Amanda was faking her emotions with me too. Throughout our conversation, her body language was rather aloof, as if she didn't really want to speak to me. Her arms were often crossed and she seemed to laugh nervously when I tried to lighten the mood. For a brief moment I worried that she regretted agreeing to the interview and that a hidden inner-Amanda was secretly harbouring ill feelings towards me. Then I realised that that would be the wrong way around; if anything, she felt nothing towards me and was just pretending to be nervous because interviews (including journalistic ones) are known to be nerve-wracking. When I told her that I hoped I wasn't making her nervous, she replied, 'If you are, I wouldn't know, so don't worry about it.'

I also wondered if Amanda remembered emotion. While we can all remember things without evoking feelings as intense as those experienced at the time, it's very hard to remember something and feel nothing. But according to Amanda, even her memories now lack emotion. 'The doctors said the damage to the amygdala was quite bad,' she said. 'Some memories of emotions might come back – you know better than me that the brain can re-grow lost connections – but I'm not hopeful.'

Her comments reminded me of another important feature of the amygdala: it generates its own kind of memory – emotional memory. Unlike the hippocampus, a memory centre responsible for the raw details of memory (what happened yesterday, where we went on a first date, etc.), the amygdala controls the emotional labelling of memory. For example, if you run into someone who was once very rude to you, your hippocampus will remember their face but your amygdala will remind you that they are a jerk. Equally, while the hippocampus's memory will direct you back to the restaurant where you and your lover shared your first meal, the amygdala's memory will infuse the occasion with warmth and intimacy. This special kind of memory is the reason we are more likely to remember things that have a strong emotional content. Even Alzheimer's patients retain more personal memories than factual ones.

Approximately 6 per cent of the general population has some form of alexithymia, a term coined by psychologists in 1973 which literally means 'no words for emotions'. It falls on a spectrum, with mild forms denoting people who experience certain emotions without understanding what they are actually feeling (something many of us are familiar with), and severe forms manifesting in symptoms such as Amanda's. For most people it's unclear what causes it. Some believe a smaller amygdala or dysfunctional neural pathways connected to the amygdala may be responsible. Numerous brain-imaging studies have shown reduced activation in the amygdala when alexithymics attempt to feel another person's feelings, recall their own emotional experiences, or imagine a happy future event. Others believe that the emotions themselves are intact, but that the brain cannot interpret them correctly; an alexithymic may for example experience excitement due to a fast heartbeat, but interpret this as anxiety and depression. Normally the brain detects changes in the body – such as tense muscles, a growling stomach, or altered blood flow – using a lesser-known sense called interoception. When we interocept, we associate a physical change with a particular emotion. Blood leaving the face is associated with fear; clenched fists are associated with anger. For reasons that remain unclear, people with alexithymia are either failing to interpret these bodily changes, or simply aren't activating the bodily changes necessary to experience the emotion correctly.

I asked Amanda what she thought lay behind her mysterious condition. Her first-hand description was intriguing.

> From everything I've read, and from my own personal experience with this thing, I think it definitely has a lot to do with how I now interpret emotion. I went to a friend's wedding a few years ago and knew I was supposed to feel joy, but instead the whole thing just felt mechanical, staged, like I was watching a movie. I don't know what I was feeling, but it certainly wasn't happiness. The weird thing, though, is that I'm pretty sure my body *was* feeling happy, because I felt a rush, a flutter, as if blood was coursing through my veins, trying to tell me, 'this is joy'. It might sound strange, but I think emotions aren't really made in the brain; they're interpreted there. Predicted, maybe.

What Amanda describes is a way of understanding emotions that dates back to the American psychologist William James (1842–1910) and the Danish physician Carl Lange (1834–1900), who revealed that the underlying physiology of emotions is not at all what it was assumed to be. Consider, they said, encountering a bear in the woods. Conventional wisdom suggests that seeing the bear triggers a feeling of fear, which in turn causes the surge of adrenalin our body needs to fight or flee. But this is actually the reverse of what happens. The truth is that when we see the bear, our body instantly responds – blood pressure rises, pulse rate increases, respiration quickens – and *because* of these physiological changes we feel afraid. The emotion arises at the end of the sequence, not the beginning. It is merely the perception of change. As James wrote,

> Common sense says we lose our fortune, are sorry and weep; we meet a bear, are frightened and run; we are insulted by a rival, are angry and strike . . . the more rational statement is that we feel sorry because we cry, angry because we strike, afraid because we tremble.[3]

The James–Lange theory remains an important idea in modern psychotherapy. Patients suffering anxiety and depression report palpable benefits after changing their body state: by performing relaxation techniques such as deep breathing, massage, tai chi, yoga and meditation, they can rein in their negative emotions and begin to feel healthy again. But arguably the theory's most significant accomplishment was that, in the spirit of Darwinism and clear-thinking empiricism, it placed biology front and centre.

The conversation soon became light-hearted. Amanda joked that she's now a kind of real-life Mr Spock, cool and detached, logical and level-headed:

> I know that's an optimistic way of looking at it, to say the least. But if I think too seriously about it my body reacts badly. I start sweating, my muscles tense, I get headaches. The doctors think it's my body telling me I'm angry. Whatever it is, it's an unpleasant sensation, so I try to avoid it.

Amanda clearly understands emotions in the abstract: she sees them through a glass, darkly – yet they remain just beyond her reach. By losing her ability to experience emotions, Amanda has inadvertently shown just how mysterious and powerful and deep-rooted emotions really are.

A Brief History of Emotion

In his classic 1872 work, *Emotions in Man and Animals*, published more than a dozen years after his landmark *Origin of Species*, Charles Darwin went beyond natural selection to explore how emotions too might have evolved.[4]

For this work, his focus was not the finches on the Galápagos Islands but the domestic dogs in his own back garden. The Darwin family loved dogs. They had a retriever (Bob), a Pomeranian (Snow), a Scottish deerhound (Bran), five terriers (Nina, Spark, Pincher, Sheila, Polly) and several large hunting dogs. Like most dog owners, Darwin enjoyed analysing his dogs' moods and behaviours. He was convinced they were analogous to our own. Some dogs, for instance, seemed jealous when he devoted attention to another dog. Others displayed hints of shame and pride, even a sense of humour. Every impression the dogs made on Darwin's endlessly curious mind was fastidiously jotted down. At one point he even described his dogs grinning. '[T]he upper lip during the act of grinning is retracted as in snarling, so that the canines are exposed, and the ears are drawn backwards; but the general appearance of the animal clearly shows that anger is not felt.'[5]

These observations led Darwin to propose a radical way of thinking about emotions – that they evolved, just like everything else, to help an animal survive and reproduce. It was an intuitive, highly persuasive idea, as blasphemous as his theory of natural selection – emotions were at the time viewed as immaterial 'passions' endowed by God – and as compelling as his secular view of all life. The message was clear: emotions are innate, based on millions of years of human evolution, and should therefore be studied from a predominantly biological perspective.

The great neurologist Sigmund Freud (1856–1939) expanded on this idea by making a clear distinction between emotions and feelings. Emotions are unconscious, he claimed, repressed by their human masters but perpetually working backstage to modify our behaviour and regulate survival. Feelings, on the other hand, are the conscious experience of emotions, a subjective phenomenon coloured by our daily lives, memories and beliefs – a story the mind constructs. Antonio Damasio, a neurologist at the University of Southern California, spent much of his career exploring this distinction and popularised it with his book *The Feeling of What Happens* (1999).

Damasio's additional insight is that feelings are also the brain's way of interpreting emotions to guide wise decision-making – vividly demonstrated by his patient 'Eliot', whose brain tumour left him unable to feel emotions. 'I never saw a tinge of emotion in my many hours of conversation with him,' Damasio recalled. 'No sadness, no impatience, no frustration with my incessant and repetitious questioning.'[6] Aloof and disconnected, Eliot lost the capacity to make decisions. His career and personal life fell apart and he ended up in his brother's care. Emotions, it turned out, really do have a mind of their own.

But the biological understanding of emotions is only half the story. What is missing from the Darwinian narrative is our own influence on emotions. We are not merely the subjects of inexorable biological forces, but rather have the power to shape and control our own emotions. As Joseph LeDoux, director of the Emotional Brain Institute at New York University, observes: 'Emotions are also a cultural and social construct. They are something our brains assemble depending on our upbringing and the kind of situation we are in.'[7] Indeed, if a human was raised in solitary confinement, with no social interaction whatsoever, she wouldn't have emotions as we know them today. She would certainly experience raw feelings, that is, bodily sensations that we call affect. And this would tell her if something was pleasant or unpleasant, unsettling or reassuring. But emotions as they exist today go beyond affect.

Take fear, for instance. We have more than thirty words in English for fear – panic, terror, worry, anxiety, trepidation, and so on. In addition, we all know that the fear of falling off a cliff is different from the fear of a job interview: one is a natural trigger, the other a socially constructed trigger. Both rely on evolved neural circuits in the brain, but the constructed trigger is ultimately the product of our environment, because we've been conditioned to think that job interviews are scary.

The truth is, we construct emotions all the time. We are not born with them. They are not hardwired into the brain. When we think about that upcoming exam, or desire someone else's possessions, or bump into an old flame, the emotions of anxiety, envy and love can appear and then disappear automatically. They are as fleeting and ephemeral as memory – and as equally dependent on the times. Just compare how we feel about the environment and gender roles today with how we felt about them only a few decades ago. We create emotions. The brain invents them, moment to moment, to make sense of an ever-changing world. We even impart emotions to nature and inanimate objects. I'm certainly guilty of this: I refer to a wilting plant as 'sad', I say that a lashing storm is 'furious' and I think London's Shard is 'lonely'.

But Darwin was not, of course, completely wrong. Evolution still had to build the amygdala and other neurological hardware to let us experience different emotions. To separate the biology of emotions from the cultures that influence them would be like separating the tides from the sea. But if you doubt how much the environment can shape the emotional brain, consider this: the ancient Romans and Greeks did not smile, at least in the way we do. There isn't even a word for 'smile' in Latin or ancient Greek. Mary Beard, professor of classics at Cambridge University, notes:

> This is not to say that Romans never curled up the edges of their mouths in a formation that would look to us much like a smile; of course they did. But such curling did not mean very much in the range of significant social and cultural gestures in Rome.[8]

Incredible though it may seem, the smile that we know and perform for the camera today is called the Duchenne smile, after the French neurologist Guillaume Duchenne (1806–75), and only became popular following the advent of dentistry in the Middle Ages.

To many, this view of emotion sounds strange. Centuries of science have taught us that emotions and feelings are purely biological products of the brain, as innate as the gastric juices secreted by the gut. Mainstream neuroscience has long theorised that emotions are fundamentally circuits in the subcortical regions of the brain, or genes that cooperate to produce specialised neural programmes – one for each of the 'Four Fs': fighting, feeding, fleeing and fornicating. The common thread in these ideas is that emotions are produced 'bottom up' from our brains, rather than 'top down' from our social environments. In recent years, however, scientists have uncovered compelling evidence to support the social constructivist view. In 2015, Carlos Crivelli, a psychologist studying emotion in the tribespeople of Trobriand, an island off the coast of Papua New Guinea, revealed that emotions are not universally understood. He showed the tribespeople pictures of Westerners with different facial expressions and then asked them to identify the emotion.[9] When asked to identify the facial expression of fear – eyes bulging, eyebrows raised and drawn together, mouth agape, lips stretched – they didn't see a frightened face. Instead, they saw aggression. It looked 'angry', one said.

Crivelli had originally sought to challenge the work of Paul Ekman, emeritus psychologist at the University of California, San Francisco, and the world's foremost expert on facial expressions, whose own research on the people living in Papua New Guinea had convinced him that emotions are cross-cultural, and thereby universal.[10] But Ekman's method suffered from a fundamental weakness. Rather than letting the participants simply guess what the emotion could be – what psychologists call free-labelling – he asked them to choose from a list of English words for different emotions, a simpler approach known as the basic emotion method. As you've no doubt guessed, the problem with the basic emotion method is that the list of words accompanying the picture essentially forces you to select an emotion that you might not otherwise have considered.

Ekman's error becomes more conspicuous when factoring in cultural differences. 'Not all cultures understand emotions as internal mental states,' writes emotion expert Lisa Feldman Barrett:

> The Ifaluk of Micronesia consider emotions as transactions between people. To them, anger is not a feeling of rage, a scowl, a pounding fist, or a loud yelling voice, all within the skin of one person, but a situation in which two people are engaged in a script – a dance, if you will – around a common goal. In the Ifaluk view, anger does not 'live' inside either participant.[11]

Clearly, different cultures create their own emotional language. Westerners, for instance, are much more likely to express negative emotions such as anger and disgust in public than Japanese people are (which must partly explain why I find the blissful serenity of Kyoto so attractive). But it's important to remember that it is our societies shaping our emotions, rather than our emotions shaping our societies. We are reacting to society, not simply feeling in a void. Just look at the public anger towards statues of slaveholders, or the outpouring of empathy and gratitude towards NHS workers during the Covid-19 pandemic. The societies we build have an extraordinary influence on how we feel and experience the world.

And yet, our brains did not evolve as rapidly as our society. Today we ceaselessly try to regulate and even medicate our emotions, all the while forgetting that certain emotions, however undesirable, may in fact have an important evolutionary purpose.

Depression's Evolutionary Roots

One evening in late 2018, a close friend of mine left work and had a nervous breakdown. Frantic, delirious and severely depressed, he paced around his kitchen like a caged animal, hurling abuse at his partner, who tried in vain to placate him. At first we put it down to another one of his mood swings – albeit a particularly bad one – and waited to see if it passed. It didn't. Over the following months he actually got worse. The doctor said there was nothing physically wrong with him. She discussed his symptoms and asked him how he felt about his life. He said he didn't really know what was wrong, but all he wanted to do was sleep. The doctor diagnosed him with clinical depression. He left work and never went back.

That was fifteen years ago, and since then I have tried to understand what happened to my friend. He had everything a person could want from life: a loving family, a fulfilling job, good friends and excellent health. But none of it mattered. His depression made him withdraw completely from the world. Now, he sleeps for fourteen hours, spends his afternoons dozing on the couch, eats very little, says very little and rarely leaves the house. Like a butterfly returning to its chrysalis, his mind has fundamentally changed.

There are times when I look at my friend and see his depression not as an illness, but as something deeper. He's not sick in the way that someone with cancer is sick. He's not about to die, even though he sometimes doesn't feel like living. What's happening to him, whatever it truly is, has been passed down from generation to generation.

According to the World Health Organization more than 350 million people suffer from depression. It represents nearly 4.3 per cent of global disease burden and the annual cost of the disorder is almost 800 billion euros. Medically, depression is defined as a persistent feeling of sadness and can be classified as mild, moderate or severe. We're still not sure what depression does to the brain, but evidence has linked it to reduced brain volume, harmful inflammation and a disruption of the chemical balance of neuro-transmitters including dopamine and serotonin. Most perplexingly, we still don't know if depression is the cause, the result, or some mixture of both in this wide-ranging array of brain damage. Depression typically first appears in adolescence, and thus should have been eliminated by the pressures of natural selection. So why hasn't it? What evolutionary advantage might it conceivably have served? There are four possibilities – social dominance, problem solving, an immune response and/or a manifestation of modern anxiety – all of which are likely to form a piece of the puzzle.

The first possibility is that depression evolved for humans to maintain their social dominance hierarchies. All societies have their ruling elite. In chimpanzee communities each member has a rank and males are dominant over females. An alpha male leads the troop, but there are other males who sometimes challenge the alpha; several males may even form a coalition to do so. If a challenge isn't met with severe punishment, the troop loses its leader and ceases to function as a group. When a bold chimp attempts to dethrone the alpha and is viciously defeated, he loses confidence and is unlikely to try his luck again. A low mood might therefore be a way to discourage subordinates in the group from toppling authority, thereby maintaining the dominance structure.[12] What's in it for the subjugated chimp? Well, he can now spend less time fighting and more time fathering offspring.

However, the social submission theory has big caveats. For one, it doesn't explain the high rates of suicide among depressed individuals. How could an adaptation to help humans reproduce also make them want to kill themselves? For another, it doesn't explain why depression occurs over such long stretches of time, often months if not years. Depression is an umbrella term for a highly nuanced phenomenon, so perhaps this hypothesis merely illuminates one aspect of its adaptive value.

The second possibility is that depression is an adaptation that evolved in response to complex problems, and is our brain's way of telling us to stop and solve these problems. Research shows that depressed people are often highly analytical: they think intensely about their problems and are usually unable to think about anything else.[13] Viewed this way, the sufferer's indifference to everything from house-cleaning to socialising to simply staying awake is instead the brain redirecting energy in order to ruminate on an important problem that has become unbearably difficult to resolve, such as a failing relationship or a struggling business. The psychologist Lauren Alloy calls this ability 'depressive realism'.[14]

Mild depression can indeed put us in a more introspective, contemplative frame of mind. In 2007 the Australian psychologist Joseph Forgas conducted a study in which he made people sad by watching a film about death from cancer, and then asked them to make persuasive arguments for or against ideas such as raising student fees, Aboriginal land rights, or whether Australia should become a republic. Compared to happy participants, a group of independent scientists found, sad participants produced more persuasive arguments that were more detailed and deliberate.[15]

In another study, Forgas showed that even memory might be enhanced by mild depression. In a grocery shop near his office, he placed unusual items including toy figurines, a tiny tractor and a little London bus on the counter, and then asked customers leaving the shop if they had noticed the items, and if so, how many they could remember. He found that on grey and wet days, when the shop also deliberately played sombre music to dampen people's spirits further, customers were far more likely to remember what was on the counter than those who had gone in on sunny days with upbeat music playing.[16] So what is it about low spirits that makes people better debaters and makes them pay closer attention to their surroundings?

The answer lies in a receptor called 5HT1A. Pharmaceutical companies are currently trying to target 5HT1A because it appears to cause depression by mopping up the brain's serotonin, the neurochemical responsible for happiness. But 5HT1A also does something else. It helps fuel neurons involved in attention and concentration. In a brain region called the ventrolateral prefrontal cortex (VLPC), neurons fire when a person wants to focus on a particular task and avoid unnecessary distractions. Without 5HT1A, these neurons cannot function properly and begin to break down.

In the modern world's obsession with happiness (which ironically makes depression harder to tolerate) the notion that sadness could be good for us almost sounds perverse. But in our ancestors' world, it would have made perfect sense. As Jonathan Rottenberg explains in his book *The Depths: The Evolutionary Origins of the Depression Epidemic*:

> One way to appreciate why these states have enduring value is to ponder what would happen if we had no capacity for them. Just as animals with no capacity for anxiety were gobbled up by predators long ago, without the capacity for sadness, we and other animals would probably commit rash acts and repeat costly mistakes.[17]

In other words, depression might be the brain's unconscious advisor – a sophisticated mechanism that forces us to focus and attend to life's obstacles. However, this hypothesis doesn't explain suicidal depression, or the different levels of depression. So maybe something even deeper is at work.

The third possibility is that depression, including severe clinical depression, is the result of an overactive immune system. Before the advent of antibiotics in the 1940s the top causes of death were infectious agents including pneumonia, tuberculosis, influenza, typhus, malaria, measles, yellow fever, typhoid and diarrhoea – among others. Such diseases were common in the ancient world and depression may have been how our ancestors fought them off.

Though it might sound a little far-fetched, geneticists are now finding that many of the genes that increase one's risk of depression also boost the brain's immune response to infection. Among them is a gene called *NYP*, which codes for one of the most abundant proteins in the brain: a neurotransmitter known as neuropeptide Y. Normally *NYP* kills infectious agents by unleashing the brain's inflammatory response, a powerful yet short-lived measure that must be tightly controlled – too little and the bugs run rampant; too much and the brain itself gets caught in the crossfire. Think of inflammation as the fiery breath of the immune system's dragon. A perfect weapon when facing fearful odds, but not something you want to be active for very long. When *NYP* mutates, however, white blood cell count skyrockets and the immune system becomes hyperactive. Figuratively speaking, a three-headed dragon that does nothing but breathe fire is born. But here's the interesting part: people with major depressive disorder are more likely to have the mutated *NYP* gene.

'It's pretty clear that inflammation can cause depression,' says Ed Bullmore, professor of psychiatry at Cambridge University. 'The question is does inflammation drive the depression or vice versa?'[18] To answer that, Bullmore and his colleagues have begun investigating whether anti-inflammatory drugs can treat depression. Preliminary evidence suggests they can. For example, when people suffering from the inflammatory disease rheumatoid arthritis (30 per cent of whom are also depressed) are treated with anti-inflammatory drugs, they often report feeling less depressed even if their physical symptoms haven't improved. On the flip side, when people are treated with pro-inflammatory drugs for conditions such as hepatitis C, they often struggle with an unexpected knock-on depression. If it turns out that depression is indeed what causes a vigorous immune response, scientists the world over will have to drastically rethink what depression really is and why nature chose such an unpleasant phenomenon, what William Styron described as a 'grey drizzle of horror . . . a storm of murk', to protect us from harm.

This is persuasive. Unlike our world of penicillin and advanced medicine, of squeaky-clean hospital floors and the near-total eradication of diseases that nearly eradicated us, our ancestors inhabited a perilous land where more than 30 per cent of the population died from infectious agents before the age of five. So it makes sense that evolution would select for such an extreme measure. Depression forces us into a kind of social hibernation, much like the self-imposed hibernation we go through when unwell, and this would have allowed our ancestors to conserve the energy necessary to fight off an infection, as well as reducing their chances of being infected by something else.

The theory also helps explain the extended duration and diversity of depression's symptoms – because infections can last a long time and are becoming increasingly diverse themselves. With this radical new approach to depression, Bullmore writes,

> We can move on from the old polarised view of depression as all in the mind or all in the brain to see it as rooted also in the body; to see depression instead as a response of the whole organism or human self to the challenges of survival in a hostile world.[19]

But what happens when the challenges of survival become finding a job, paying the rent, or other distinctly modern incarnations?

The fourth possibility is that depression is a form of modern anxiety. Kierkegaard described anxiety as the dizziness of freedom, and W. H. Auden warned that freedom, especially in the modern world, would trigger an age of anxiety. Both thinkers had arrived at a fundamental truth about our world. We are uncomfortable with choice. From cuisines to mobile apps to professions and life partners, we are almost paralysed by the number of options available. And this can lead to a collective angst that might be the reason why depression is on the rise worldwide.

Depression is rare in agricultural societies. In a study of 2,000 Kaluli tribespeople from the rain forests of Papua New Guinea, only one case of clinical depression was found.[20] The lifestyle of the Kaluli is probably very similar to the one early humans experienced. Their work and social options are limited and they cannot venture too far from home. To some this might sound like a less fulfilling life, but when the basic needs of being human are met – food, shelter, love – life can foster a significance and contentment many alive today will never realise. It is an inescapable fact that civilisation can be a burden, an exhausting quest for identity and meaning in a world where the boundaries for such things are increasingly blurred. The once homeless and crack-addicted author Lee Stringer wrote in his essay 'Fading to Grey':

> Perhaps what we call depression isn't really a disorder at all but, like physical pain, an alarm of sorts, alerting us that something is undoubtedly wrong; that perhaps it is time to stop, take a time-out, take as long as it takes, and attend to the unaddressed business of filling our souls.[21]

I can relate. There was a time, during my early thirties, when I had something of an existential crisis. I had split up with a partner I had been with for six years. I was devastated and took a long time to move on. In the midst of my sorrow, I travelled to America to visit some friends. I thought the distraction would act as a sedative. Plus, despite not being in the best frame of mind, I was eager to meet new people and explore where life could take me. But as time went by – as I travelled through New England, the South and the Californian coast, soaking up the richness of my surroundings and listening to friends wax lyrical about why their state was the place to be – I realised something I'm still a little embarrassed to admit. I wanted it all. The skyline of Manhattan was dazzling; the hospitality of Nashville life-affirming; the beauty of Los Angeles overwhelming. To choose one life was to sacrifice the rest. I didn't know what to do. It was dizzying. I became anxious and even more depressed.

You might think that I am merely spoilt; I should have just been grateful for the wealth of choice that globalisation and modernity provided. Perhaps. But speak to any psychologist and they will tell you that too much choice is making us more regretful, indecisive and unhappy. As social psychologist Hazel Rose Markus notes,

> We cannot assume that choice, as understood by educated, affluent Westerners, is a universal aspiration, and that the provision of choice will necessarily foster freedom and well-being. The enormous opportunity for growth and self-advancement that flows from unlimited freedom of choice may diminish rather than enhance subjective well-being.[22]

The daily routines of modern life are also taking their toll. Just look at our sleeping habits. In England and America, countries where depression rates are among the highest in the world, the average amount of sleep fell from nine hours in 1910 to 6.8 hours in 2016. Our cult of productivity exposes us to less natural and more artificial light, both of which are linked to depression.

The good news is that governments are attempting to solve the problem. By tapping into our emotional brains' evolutionary potential, the Nordic countries in particular have had great success. Leading the way is Finland, which ranked number one on the UN's 2021 *World Happiness Report.*

The Social Construction of Emotions

The weather in Finland leaves much to be desired. Northern winds foster dark, arctic winters; the summers are short, typically two months or less; and due to the country's ill-fated latitude, some parts of it receive almost no sunlight at all. Hardly cause for jubilation and high spirits. Yet by nearly every metric of happiness Finland comes out on top. The Finns enjoy the highest levels of income, life expectancy, social security, freedom of choice, freedom from corruption, generosity, trust, faith in the police, and immigrant well-being: 'a Finnish miracle', the Helsinki-born economist Bengt Holmström told the press. A bastion of socially progressive values, Finland also boasts the third most gender-equal society (just below Norway and Iceland) as well as astonishingly little income inequality.

Compare this with nations such as Haiti, Syria and the Central African Republic, where endemic corruption and civil war has torn people apart like a scene from the Old Testament. In South Sudan the ethnic cleansing is in danger of becoming a repeat of the 1994 Rwandan genocide, in which 800,000 people were slaughtered. And it's not just third-world countries suffering the most. The US ranked eighteenth, falling five places since 2016. According to the report, America's well-being is being 'systematically undermined by three interrelated epidemic diseases, notably obesity, substance abuse (especially opioid addiction) and depression.' This, in a country where income per capita has more than doubled since 1972. What happened? Scholars call it the Easterlin Paradox, after the economist Richard Easterlin of the University of Southern California who in 1974 declared that there is no link between a nation's economic development and the overall happiness of its citizens.

So how has Finland done it? Well, in biological terms the Finnish people are displaying the emotional brain's best-kept secret, a masterpiece of organic harmony that transformed our species from warring apes to children of the Enlightenment: the evolution of love. Like every human emotion, love grew out of necessity. As the brain grew bigger and *Homo sapiens* expanded across Africa, Europe and Asia, natural selection began to favour genes underpinning bonding, cooperation and altruism. Slowly but surely, humans realised they were better off forming large cohesive groups that watched over each other. Teamwork was rewarded. Friendships blossomed. And though many in today's tumultuous world seem to have forgotten this – indeed rampant individualism has atomised much of Western culture – the Finns certainly have not. They have a saying, a mantra they pass from one generation to the next: *talkoo*. It means working collectively for a greater good, forming communities based on non-hierarchical principles.

Here we return to the central tenet of this chapter: social constructs such as *talkoo* shape our emotions far more than we realise. 'Humans have a capacity to create social reality – that is, you and I can agree something has a function by virtue of our agreement,' Lisa Feldman Barrett told me. A good example of how humans engage in this phenomenon is money. Those little pieces of paper we use to trade for material goods have value because we agree that they do, and if we hadn't agreed that, we wouldn't use them. In the same way, if we agree that crying can be due to anger, sadness or happiness, then it becomes that. Like money, we've taken something physical – crying – and imposed a social meaning on it.

This way of thinking isn't meant to dismiss biology. Social reality wouldn't exist without the biological processes that create human minds. 'What the Finnish have managed to do,' says Barrett,

> is find a way of living where they ease the brain's energetic burden on each other. So they can free up their available energy to do creative things, to innovate and make energy investments in each other. That's what humans do. We regulate each other's nervous systems by social concepts. And the Finnish have found a way to do this that is maximally effective.

While we still don't fully understand how the brain accomplishes this extraordinary feat, most neuroscientists agree that two neurochemicals create the biological basis of emotional attachment: dopamine and oxytocin, key regulators of emotional and social behaviour, nicknamed 'the love hormones'. Both do something quite extraordinary. They deactivate the amygdala, reducing anxiety and fear, and enhancing empathy, compassion and social interaction instead. More precisely, they activate GABAergic neurons, the brain's off switches, which tell every neuron in the vicinity to stand down. With the amygdala effectively offline, dopamine and oxytocin are then free to activate neurons in the frontal cortex important for good social behaviour. The effect is striking. Give oxytocin to people with anxiety or PTSD, for instance, and their symptoms soon vanish.[23] Oxytocin is released during hugging and orgasm, and helps quell depression and stress.[24] Even at the molecular level, it seems, love erases fear and rewards cooperation.

In 2012 researchers at Bar-Ilan University in Israel measured the blood oxytocin levels of 120 new lovers (sixty couples) three months after they began their romantic relationship and then again six to nine months later.[25] They then compared these oxytocin levels with those of non-attached single people, and the difference was huge. Not only did new couples have much higher levels of oxytocin coursing through their veins, but these levels also showed no signs of decreasing during the first nine months of the relationship. The levels of oxytocin were almost as high as those observed in new parents, proving that parental and romantic attachments undoubtedly share some of the same evolutionary mechanisms.

Building on these findings, German researchers found that administering oxytocin to men in a relationship made them perceive their girlfriend's face as more attractive than another woman's. The hormone, in effect, helped keep them faithful.[26] Chemically treating the declining sexual attraction that happens to many couples might sound controversial. But as the researchers point out, after several years relationships sometimes shift from passionate love to compassionate love, similar to friendships. So would it really be that bad to rekindle the passion in a relationship, to relive the honeymoon period whenever we wanted?

Love is so powerful it is actually a kind of addiction: it uses the same brain reward circuits that hook drug addicts to cocaine and heroine. This is why, when we fall in love, the source of our desire can do no wrong. It is also the reason people risk ruining their marriages with illicit affairs; the blissful, giddy euphoria we feel when we are in love shuts down neurons in the frontal cortex responsible for judgement. Those who are persistently unfaithful in a relationship may have what's called a 'flame addiction', a constant craving for the chemical and hormonal changes in the brain that occur in the build up to a new sexual partner, an addiction solely to the chase. The main culprit in a flame addiction is dopamine. For the compulsive cheater, that feeling of reward that comes from high levels of dopamine in the brain – the sense of accomplishment – is what they are actually addicted to.

But despite love's darker sides, when the conditions of a group are fine-tuned to maximise love, as they clearly are in Finland, the effect it can have on a society is staggering. And that's really all the Finns have done. Their brains are no more evolved than ours. It's certainly possible that some genetic factors are at work (a recent study did find that Finns are an independent genetic population, not a part of the European population as was once believed).[27] But whether by nature, nurture or a peculiar coalescence of both, Finland's people are demonstrating what can be achieved when we let go of our isolated lives, our unfettered individualism, our toxic tribalism, and simply embrace the brains that evolution has given us.

Because the truth is that happiness – real, quantifiable happiness – is also a social construct. That's not to say that happiness has no biological meaning; clearly it does, as dopamine and oxytocin prove. But the fact is we have also evolved brains to be more responsive to environmental influences and opportunities. A condition in the environment that maximises emotional well-being, such as the Finnish spirit of *talkoo*, may emanate from biological beings but is nonetheless an invention of a particular culture, which can no doubt be replicated elsewhere. The Japanese have their own version, *amae*, which means loving another person and being able to depend on them like a child can depend on his or her parents. *Amae* was first described by the Japanese academic Takeo Doi in his 1971 bestselling book *The Anatomy of Dependence*, which he wrote after experiencing the culture shock of American independence (something he thought they took too far).[28]

The notion of happiness as a social construct tends to unsettle people. It conjures images of a dystopian world where thought-police tell us what we should and should not find pleasing. If happiness is unduly influenced by the social sciences, many warn, we are on a slippery slope to tyranny. After all, how is the happiness a dictator feels towards his money and power, or the happiness a religious fanatic feels towards their god, any less valid than the happiness a Finnish person feels towards camaraderie, equality and human rights? It's a fair point. But to view emotion this way is to view it in isolation, and the brain doesn't work that way.

Thus what also evolved at this stage of the brain's evolutionary tale was a natural corollary of emotion, and something that has come to dominate nearly every aspect of modern life. If you haven't already guessed, clues lie in emotions such as embarrassment, shame, jealously, and empathy – all of which are tightly linked to the development of social groups. Which, as the human population grew, is precisely what the brain needed to adapt for.

3

Our Social Brains

Chris Rees was fifteen when he first broke the law. Growing up in an impoverished high-rise in north London, Rees joined a gang and quickly fell into a life of petty crime. He's been in and out of prison (a place he says only encourages criminal behaviour) and has now been diagnosed with anti-social personality disorder (ASPD), a mental health condition that causes a person repeatedly to engage in reckless, impulsive and often violent behaviour.

'Being social wasn't an option,' Chris, now twenty-four, told me during our long conversation in London's Regent's Park, a place he said makes him feel calm and at ease with the world. 'I saw everyone as a threat, but also as a kind of opportunity. I was always just thinking about myself. I couldn't relate to other human beings.'

Chris is intelligent and charming. His worst crimes, however, include knifepoint muggings and assaulting his brother. Like other personality disorders, ASPD exists on a spectrum, with minor offences at one extreme and sociopaths at the other. Chris is somewhere in the middle. He sees a counsellor once a week and is receiving psychological therapy.

I had first heard of Chris a year earlier through a research colleague. I'd asked if she knew anyone whose brain condition could shed light on the evolution of social minds. The field was still very much in its infancy, so I wasn't entirely sure what I was looking for. But when I called Chris to arrange our meeting, I knew he could provide insight. He said he was tired of being perceived as a criminal and wanted a deeper understanding of his condition.

'Talking to someone definitely helps,' he said, as we strolled through the park. 'They tried a bunch of random medications, but they just put me to sleep. You can't be social when you're a zombie. The social part of my brain, which I guess is malfunctioning or just hasn't really developed, needs people, not pills.'

The causes of ASPD are partly social themselves: inequality, poverty, lack of social mobility, time spent in dysfunctional youth prisons, addict parents – all are strongly linked to ASPD. But research is also identifying genetic risk factors for the condition, often in genes linked to aggression and impulsivity. Variations in the gene *MAOA* (*monoamine oxidase A*), nicknamed the 'warrior gene' because it's thought to drive aggression, have been found to contribute to ASPD in maltreated children.[1] Other candidate genes including *COL25A1* (*collagen type 25 alpha 1 chain*), *CDH13* (*cadherin 13*) and *LINC00951* (*long intergenic non-protein coding RNA 951*) have been linked to ASPD, though their mechanism of action remains unknown.[2] Chris considered having his genome sequenced to learn if he possesses these genes, but said he isn't ready for a psychological battle with genetic determinism.

> I know genes aren't destiny – at least, that's what I read – but I still feel like knowing that kind of information would set me back in my struggle with this thing. I want to get better. I need to get better. And I think understanding my brain and how I got here will get me there faster than just saying, 'Oh well, you've got these bad genes, but we have no idea what they do.'

In the brain, ASPD wreaks havoc on the ventromedial prefrontal cortex (vmPFC), which controls social cognition and the perception of threats. While little is known about how the vmPFC performs these tasks, we know that neural networks within this region are essential, and acutely sensitive to change. Damage one neural network in the vmPFC and a person may never be able to learn from their mistakes; damage another and a person may be easily misled and unable to make their own decisions (something Chris said his old gang exploited ruthlessly in him). As part of the cortex, the brain region that expanded to occupy 80 per cent of the human brain (compared to only 28 per cent in rats), the vmPFC represented a critical evolutionary leap. The vmPFC is very ancient, and probably evolved in apes more than 15 million years ago.

When I told Chris that I was writing a book about brain evolution and that I wanted to understand what his condition tells us about our social brains, his response was astonishing:

> We hear about evolution a lot – it was definitely drummed into my head at school – but we still don't think of humans as a messy, imperfect result of it. We still think we're above it. Separate from it. Well, we're not, and I'm sorry if my brain doesn't work the way it should, but nature's been changing human brains for how long?

'Seven million years,' I said.

'Jesus. Seven million years. How are any of us supposed to be "normal" after that?'

As unfortunate as Chris's situation is, his condition provides testimony to this ancient brain history. For we are all only as social as evolution has allowed us to be. One of the greatest accomplishments of our evolving minds is their ability to connect with other conscious beings. Be it through family or friendship, bonding at work, team sports or religious affiliation, *Homo sapiens* are known for manifesting intense social relationships not seen in any other primate including our early-human predecessors. Only humans form complex communities in which an outsider is welcomed, given food and shelter, and allowed to integrate for the benefit of the entire group. Only humans will allow someone completely unrelated to hold their new-born babies. Only humans organise themselves into networks capable of such staggering feats as sequencing their own genome and building the Large Hadron Collider. Quite a difference, given that we share 98 per cent of our DNA with apes who would fight to the death to avoid carrying a log together.

To understand how our social brains came about, we must again travel back to East Africa more than 4 million years ago.

The Origins of Sociability

Our human forebear Ardipithecus holds many of the clues to understanding our social brain.[3] Standing four feet tall, with a strikingly large toe to grasp branches and long curving fingers to climb trees, she was a forest-dwelling primate who developed social thinking primarily to avoid predators and gather food. She preferred wooded or forested habitats, where she lived alongside leaf-eating monkeys, parrots and peacocks. Fossils from Aramis in the Middle Awash region of Ethiopia suggest that she lived in groups at least thirty strong. Fossils of her teeth suggest she was a fruit eater, rather than a plant and leaf eater, and studies show that apes tend to forage for fruit as a cohesive social unit, relying on each other to find fruit in the sporadically distributed fruit-bearing trees, mark its location and remember when it will ripen.

The logic of social harmony is simple enough: if a band of hunter-gatherers (at first a single family which then expanded into a group of several families) works together to outwit predators, everybody eats. The loner who eschews this strategy isn't going to last long, presumably taking his anti-social genes with him. By working together towards the same goal – what biologists call mutualistic cooperation – humans were able to rise above their selfish impulses and in turn create new methods for hunting, fishing, foraging and threat avoidance.

These practices were then passed to future generations using basic gesturing. A member of the group would point to the treetop harbouring a predator, or to a lone gazelle sheltering in the forest glade. Only humans, domestic dogs and, perhaps, elephants understand the unspoken signal of pointing; chimps do not. And though pointing may sound primitive, this type of gesture laid the foundation for abstract thinking: it signified the ability to think about something other than one's self. Where once humans thought solely in terms of individualistic needs and desires – a ruthless dog-eat-dog struggle for survival – now they understood the power of good collaboration based on the concept of shared intentionality. Where once humans understood only themselves, now they understood the power of the collective.

Before long an entire culture based on these instincts was born, and social minds eventually mushroomed into armies, religions, governments, free markets and a worldwide matrix of social media. Needless to say, to study something so entwined with the social norms and cultural conventions of all of human history is an enormous challenge. Fortunately, humans alive today provide excellent clues. In addition to stories of patients such as Chris Rees, we can use advanced neuroimaging techniques to observe brain activity during a particular task. The neuroimaging I'm most familiar with is functional magnetic resonance imaging (fMRI). Based on conventional MRI, which uses a strong magnetic field to create detailed images of the body, fMRI goes a step further by looking at blood-flow changes in the brain. Such changes then tell us – or at least strongly indicate – which areas of the brain are most active in response to different stimuli, because blood flow and brain cell activity are intimately linked. Today, fMRI is used to investigate vital functions such as speech, attention, vision, touch, memory, pain and emotion. When I was doing research into Alzheimer's disease, I often used fMRI data to see how the brain stores and retrieves memory, the results usually showing the brain's hippocampus and cortex lighting up like a Christmas tree. Like all neuroscience techniques, fMRI has its limitations: the resolution is low and it can only be used on a single person remaining perfectly still. But it is still the only technique that offers a tantalising glimpse into the human brain, and is the closest we can come to watching the brain at work.

In the context of the social brain, fMRI studies show that our ability to connect with others depends on a network of 'neural modules', a constellation of circuits that links the amygdala to the parietal and prefrontal cortex. There is no single site controlling social interaction; rather, each module is activated depending on the social activity we engage in. For example, modules in the prefrontal cortex are activated when we have a conversation with someone about something we are both familiar with; modules in the amygdala are activated when we are pondering whether we like someone; and modules in the parietal lobe are activated when we simply observe someone's body language. As Matthew Lieberman observes in his remarkable book *Social: Why Our Brains Are Wired to Connect*,

> Just as there are multiple social networks on the internet such as Facebook and Twitter, each with its own strengths, there are also multiple social networks in our brains, sets of brain regions that work together to promote our social wellbeing. These networks each have their own strengths, and they have emerged at different points in our evolutionary history moving from vertebrates to mammals to primates to us, *Homo sapiens*.[4]

To understand how the first social brain operated, let's rewind the clock and imagine performing an fMRI on our 4-million-year-old ancestor Ardipithecus. Granted, she'd likely be somewhat displeased by the prospect and, despite being gentler than a chimpanzee, would probably try to kill her *Homo sapiens* experimenters. But let's pretend our Ardipithecus ancestor plays along. Ardipithecus had a small brain (about 400–550 cm^3), similar to that of a gorilla and about a third of the size of *Homo sapiens*, but one that nonetheless contained sophisticated social brain circuitry. She possessed a cerebral cortex, albeit a shrunken version of ours, which probably stored microcircuits of neurons with a similar basic structure to our own. Our fMRI scan might, therefore, detect a flash of activity in Ardipithecus's frontal cortex during an activity requiring social cognition, such as seeing pictures of her kin. She also possessed an amygdala, which would have strengthened her social ties by imbuing them with emotions such as love, happiness and fear. This region of Ardipithecus's brain would also light up in our fMRI machine in response to the sound of her offspring in distress. But the connections between Ardipithecus's social brain networks were not as advanced as ours, making each flash of brain activity a relatively isolated event.

One way to understand how these social networks evolved is to think about the earth's natural history. When continents drift across the ocean bed they produce a constellation of landmasses, each unique and each containing enough geological evidence to reveal how they moved over time. So it is with the social circuitry of the brain. Here, however, landmasses can be represented by the neural circuitry for a variety of social behaviours, which neuroscientists can then use as building blocks to understand how human social circuits evolved over time.

To bring this to life, consider the cetaceans, a group made up of dolphins, whales and porpoises which evolved from land-living mammals that re-entered the ocean over 50 million years ago. They have astonishingly rich social lives – hunting together, babysitting one another's young and cooperating with other species (famously helping Brazilian fisherman with their catches). This appears to be due to the expansion of the cetaceans' brains, an evolutionary process known as encephalisation. This suggests that social circuits were once buried deep within the brain, bunched together like the innards of a golf ball. Then, like the Pangaea supercontinent breaking apart, the social brain began to expand, spreading its cellular tentacles into every corner of the brain, establishing new connections to oversee every aspect of social life.

We still don't know precisely how social genes and the social environment came together to shape the social brain, but research indicates that a person's genes and their environment interact to influence their social lives.[5] No surprises there, you may think. But you would be surprised at how many people still contend that one is more important than the other. To those who pitch their tent on only one side of this debate, I am afraid you have two incontrovertible facts to confront. 1. A person raised by gay parents in San Francisco is going to have a radically different social attitude towards homosexuality than a person raised by members of the Taliban in Afghanistan and Pakistan. Ergo: nurture matters. And 2. Although *Homo sapiens* crammed together on an airplane can suppress any feelings of irritation and behave altruistically, if replaced by a planeload of chimpanzees, as primatologist Sarah Hrdy has pointed out, 'Bloody earlobes and other appendages would litter the aisles.'[6] Ergo: nature matters as well.

Among the first to recognise our brain's susceptibility to the social environment was the sociologist Gustav LeBon, who published a book in 1885 called *The Crowd: A Study of the Popular Mind*. Inspired by Darwin's zeal and intrepidness, LeBon travelled throughout Europe and Asia to study the different peoples and civilisations of the world, eventually proposing an idea that he called the 'crowd mind'. He argued that human behaviour is largely guided by external influences and that every person's mind is merely 'a grain of sand amid other grains of sand, which the wind stirs up at will.'

Both Hitler and Lenin are known to have read *The Crowd*. They knew only too well that people could ignore their sense of right and wrong to be included in the tribe; that people often behaved irrationally to fit it. LeBon knew it too. In 1871, at the age of thirty, he witnessed the Paris Commune when, for two months, crowds of far-left extremists set fire to buildings and staged executions by firing squad. They killed more than 15,000 people targeting farmers, workers, priests, the rich and social democrats. The

French poet Anatole France, himself a socialist, described it as 'a government of crime and madness'. The crowd mentality, LeBon came to realise, was a terrifyingly fragile ecosystem.

Among the first to show how genes can influence our social brains was the honeybee (*Apis mellifera*). Though most types of bee are solitary, honeybees are decidedly social insects – living together in large, complex societies consisting of a single queen, hundreds of male drones and anything from 10,000 to 80,000 female worker bees. They're so social, in fact, that they're referred to as 'eusocial': an advanced level of social existence in which the animal lives in multigenerational family groups, with one animal (in this case the queen) reproducing, and the others working to care for the next generations. They communicate using elaborate social signals including the waggle dance, round dance and shaking signal, as well as touching antennae (antennating) and exchanging food (trophallaxis). When the honeybee genome was finally sequenced in 2006, *Nature* ran a cover declaring, 'A blueprint for sociality'.

By all appearances, honeybees and humans are radically different organisms, yet we share 44 per cent of our DNA with honeybees. And when viewed as a single superorganism, honeybees behave in the same way that neurons in the human brain react to the outside world, obeying psychophysical laws linked to perception and our response to external stimuli. For this reason, researchers are investigating the honeybee in search of shared social genes. One that stands out is an ancient gene called *Fushi Tarazu Factor 1* (*FTZ-F1*), which codes for a nuclear receptor: a molecule that binds DNA to regulate the expression of multiple genes. Another is a set of genes encoding a protein called heat-shock protein 90 (HSP90), which helps stabilise and fold other proteins in the cell. While we don't yet know how these molecules regulate social behaviour, the overall conclusion is inescapable: our social brains and the social brains of other animals, especially the honeybee, share deep evolutionary roots.

Over time evolution enhanced the connections between the brain's social circuits, allowing the brain to adapt to whatever social and cultural milieu it found itself in. As more complex social systems emerged in the world, the brain developed neural circuitry in the temporoparietal neocortex: a region crucial for understanding people's intentions, beliefs and personality traits. A million years later, this circuitry gave rise to empathy and the realisation that others may have different perspectives from our own.

Empathy

Darwin, for his part, was fully persuaded that empathy lay at the heart of human sociability. In another of his lesser-known works, *The Descent of Man*, a brilliant and trailblazing piece of evolutionary psychology published in 1871, eleven years before his death, he did a U-turn on the notion that humans are opportunistic, competitive and ultimately selfish animals. Instead, he declared, we humans are naturally equipped with a compassion and empathy (derived from the German *Einfühlung*, 'feeling into') that often values the welfare of others – including nonhumans – more than our own survival and reproduction.[7] Why else would humans engage in the biologically senseless act of altruism, doing things for others with no guarantee of reciprocity?

The primatologist Frans de Waal believes that primates are uniquely predisposed to empathy. He has shown that when two monkeys are placed side by side and one monkey is offered one of two tokens – one that rewards only itself with a slice of apple, the other that rewards its partner also – the monkey nearly always opts for the pro-social option, even when paired with a monkey he or she has never come across.[8] De Waal adds that fear doesn't come into the equation because dominant monkeys are in fact the most charitable.

Empathy is hard to spot in the brain. Just like the neural hardware governing social interaction, the brain has not evolved with a specific empathy centre or empathy neurons. What appears to have evolved instead is a distinct network of cells encompassing the temporoparietal junction (which lets us think about others), the dorsolateral prefrontal cortex (which lets us think about others' wellbeing), and the orbital and ventromedial cortex (which lets us determine an appropriate response). In humans, the ability to intuit what another person is thinking and feeling develops early, at a point that psychologists call the ninth-month revolution. From then on infants will look where their parents point and follow their gaze, offering the first signs of an emotional and social connection. Avoiding eye contact at this stage can be a symptom of autism: a different kind of mind characterised by differences in social interaction. The National Institutes for Health describes autistic children as 'indifferent to social engagement', stating that some 'only interact with others to achieve specific goals.' Yet as we will see in chapter 8, autism has played an important role in brain evolution and there are fascinating evolutionary reasons for its existence.

The ninth-month revolution is only the beginning. Well into adulthood human minds use culture to understand others' states of mind. Studies show, for instance, that children living in Australia and the United States understand that people have different opinions on a subject before they understand that some people are simply ignorant about a subject; but for children growing up in China and Iran, it is the other way round.[9] In Samoa, where a culture based on Fa'a Samoa ('the Samoan way') considers it taboo to question the truthfulness of local beliefs, children understand what false beliefs are five years later than children raised in Europe and North America.[10] Japanese children recognise the importance of criticism more than Italian children.[11] Fijian children appreciate the value of experience over education more than children living in the West.[12] Everywhere one cares to look, cultural practices wire our social brains in astonishingly diverse ways. And these differences appear to depend largely on what a culture does and does not decide to talk about.

These findings suggest that social brains evolved at least twice – first as a way to recognise another's feelings through empathy, second as a way to share another's feelings through culture – and that both adaptations were critical to the survival of our species. But while those traits were one source for the evolution of social minds, another was even more surprising.

Fair Play

Rats like to play. Place two in a cage and they will rough and tumble like there's no tomorrow. Place two differently sized rats in a cage and the same thing happens, only the larger rat will assert its dominance by winning the play fight. Given the opportunity, the large rat will do this repeatedly. But let the experiment run for long enough, as many neuroscientists have, and something interesting happens. If the larger rat doesn't let the smaller rat win at least 30 per cent of the time, the small rat no longer wants to play. No one enjoys a game they constantly lose. It's just not fair.

Likewise, when humans play an ultimatum game in which one player decides to distribute £10 between himself and another unevenly – keeping, say, £9 for himself while giving just £1 to the second player – the second player almost always rejects the offer despite being worse off as a result. One might think it's better to get something instead of nothing, but this doesn't play out in human interactions. We prize fairness over material gain. The question is: why?

Fair play probably evolved in the social brain to balance power. Vampire bats are a brilliant example. When a hungry bat fails to suck the blood of its prey (usually other mammals but occasionally humans), it will return to its cave and beg one of its fellow bats to share from their own bloody spoils. Remarkably, the request is granted and the fellow bat regurgitates some of its meal of blood for the hungry bat to devour. Because the bats remember which bats have shared blood with them in the past, they reciprocate if the charitable bat one day finds itself struggling for food. The behaviour provides hunger insurance as well as a sense of camaraderie. Cleaner fish and their hosts (or 'clients') are another example of this. In exchange for chasing off possible dangers to the cleaner fish (including not eating it itself), the client gets its ectoparasites removed, which if left can cause serious injury. Most cleaners, however, prefer to eat the client's mucus or healthy skin and will do so if they can. They cheat, in other words. The client then has to decide if the cleaner fishes' services outweigh the occasional act of cheating. Sometimes a cleaner fish will gently rub itself against a client to gain trust and form a memorable bond, signifying that it wants to play fair.

Unsurprisingly, evidence of fair play in humans is harder to find. At the time of writing, the richest eight people own half the world's wealth. Barack Obama called this 'the defining challenge of our time'. Yet the hunter-gatherer societies that have survived into the twentieth century suggest that we have evolved the brain hardware for fair play. The Kalarahi bushmen, African pygmies, Andaman islanders, Greenland eskimos, Australian aborigines, Paraguayan Indians and Siberian nomads all behave in an egalitarian manner. Food is shared among the tribe and members are treated equally. And while individual autonomy is encouraged, a strong emphasis is placed on collaboration and humility. When the anthropologist Richard Lee asked the indigenous people of Botswana about a practice they call 'insulting the meat', an elder replied,

> When a young man kills much meat, he comes to think of himself as a big man, and he thinks of the rest of us as his inferiors. We can't accept this. We refuse one who boasts, for someday his pride will make him kill somebody. So we always speak of his meat as worthless. In this way we cool his heart and make him gentle.[13]

Fair play suppresses the desire to dominate. It's what stops the large rat tearing the small rat to shreds. It's what keeps the bat roost committed to distributive justice. It's what stabilises the human conflict between individual and collective thinking.

The reason human brains have evolved a fondness for inequality is because our minds intuitively draw a distinction between unfair equality (all students receiving the same exam grades regardless of merit) and fair inequality (the doctor earning more than the cleaner). When push comes to shove, humans nearly always prefer fair inequality to unfair equality. This is what allowed us to work together in large groups. As the cognitive scientist Mark Sheskin notes:

> Wouldn't you prefer to team up with someone who puts in at least a fair share of the effort and takes at most a fair share of the reward, rather than somebody who is lazy or greedy? Likewise, others will prefer to interact with you if you have a reputation for fairness. Over our evolutionary history, individuals who cooperated fairly outcompeted those who didn't, and so evolution produced our modern, moral brains, with their focus on fairness.[14]

What happens in the brain when we engage in fair play may surprise you. It produces a surge of neural activity in our reward centres, a set of brain structures called the striatum, the ventral tegmentum and the ventromedial prefrontal cortex. When the brain is viewed through neuroimaging, these areas light up when money is distributed evenly among participants in an ultimatum game. They light up during friendly behaviour in so-called trust games. They even light up if the winner of a cash prize sees another person win a prize that reduces the value of her prize. In other words, fairness feels good. When the brain's reward centres are activated by fair play, they release a potent cocktail of chemicals that influence our happiness including dopamine, oxytocin, serotonin and endorphins. Individually, these chemicals offer a slight buzz. But together, they can make us ecstatic.

In his book *The Expanding Circle*, the philosopher Peter Singer argues that *Homo sapiens* are gradually widening their circles of compassion to include all people and all sentient life.[15] I think he's right. Humans are increasingly participating in what scholars call the 'humanitarian revolution': an astonishing decline in violence in both the long and short term. While social brains began their evolutionary journey by connecting our biological kin, they are now connecting all social groups. They have taken what was once purely genetic and transformed it into a consciously chosen ethic. Despite all the doom and gloom about social media and how it's making us lonelier, our brains are evolving to be more social, not less. Nevertheless, there may be a limit to how social we can be.

Social Networks

In the late 1980s a British anthropologist named Robin Dunbar noticed a curious relationship between the size of a primate's brain and the size of its social group – the bigger the brain, the bigger the social group.[16] He spotted this after measuring the size of the orbitofrontal cortex (a region at the front of the brain involved in higher thinking and advanced cognitive functions) in monkeys and apes. He found that it correlated with a variety of social behaviours: a large orbitofrontal cortex was linked to more social play and social learning, more elaborate grooming networks, higher rates of deception and coalition, a greater number of females in the group, and even an increase in male mating strategies. This led to a spellbinding new theory of human brain evolution: the social brain hypothesis.

Simply put, it states that humans need large brains to manage their remarkably complex social systems. However, there is a constraint on the number of individuals a person can maintain a stable relationship with, which Dunbar calculates to be about 150 people – Dunbar's number, otherwise known as a 'clan'. It turns out that a striking number of human organisations from factories to villages to armies operate around units of about 150 people. And the vast majority of Facebook users list around 150 friends.

On the surface, Dunbar's hypothesis did seem to reflect a truth about the social brain. Other brainy mammals, such as sperm whales – which have a brain six times larger than that of humans – display unusually sophisticated social behaviours. They travel in pods 500-strong, avidly fool around and play with one another, and communicate using a fascinating pattern of clicks called 'codas', some of which are believed to represent names of particular individuals. Conversely, many small-brain animals (owls, foxes, koalas, sea turtles) spend the majority of their lives completely alone. Moreover, archaeologists now believe that Neanderthals disappeared not because of their lack of intelligence but because they tended to be anti-social, limiting themselves to groups of twenty to fifty people in small territories while *Homo sapiens* travelled 100 kilometres or more to find new friends.[17]

Also important is the *shape* of one's social network – that is, whether your friends know each other directly or through you alone. Ronald Burt, a sociologist at the University of Chicago who studies how social networks create advantages in the workplace, is particularly interested in the latter shape. In the realm of sociology research, those who befriend people who wouldn't otherwise know each other are called information brokers. In 2004 Burt wondered whether information brokers might possess an edge over people who only connect with members of their own group. To find out, he and his colleagues looked at data describing 673 managers in the supply chain of Raytheon, America's largest electronics company based in Waltham, Massachusetts. Each manager was asked to write down how they think business performance could best be improved. Executives at the company then rated their ideas.

The most valuable ideas, Burt found, came from managers who connected with people outside their work group – the information brokers. They offered smarter solutions to problems and tended to be promoted more often, earning higher wages as a result.[18] 'People who live in the intersection of social worlds are at higher risk of having good ideas,' Burt told the *New York Times.* Bridging social worlds gives brokers what he calls a 'vision advantage'. Where others see limited options for handling difficult situations, they see fresh perspectives and divergent interpretations. Where others restrict themselves to the narrow opinions, phrasing and behaviour of a single group, they branch out and open themselves up to unorthodox views and novel ways of thinking.

Neuroscientists think that this kind of behaviour changes the way the brain works: that the brain evolved both to influence and *be* influenced by the social circles we keep. When we look more closely at the regions being used, we see that they overlap with what psychologists call the mentalising network: a dense labyrinth of neurons and fibres crucial for understanding other people's intentions, beliefs and desires.[19] Researchers studying this network are particularly interested in von Economo neurons, a strange type of brain cell only found in whales, elephants and apes – including humans. We believe they evolved in highly gregarious animals specifically to deal with complex social behaviours.

Of course, the size and shape of a group are not the only factors determining the evolution of social minds. A flock of birds or a swarm of bees is a large group taking various forms, yet still little more than a cluster of individuals all doing the same thing, a hive mind built from a few simple social interactions. 'If it were only the size of social groups that mattered, wildebeests would be wizards,' declared the social anthropologist Joan Silk.[20] What makes human brains special is that they have evolved to handle complex social situations within groups as well. And this arises from a deeper human necessity, a social contract written into the fabric of our DNA that set all forms of social interaction in motion. It is a phenomenon universal to all human societies, but we call it the nuclear family.

It Takes Two

Biologically speaking, monogamy (pairing for life) is hard work. The monogamous brain consumes more energy because it has to consider a partner's perspective as well; for our ancestors, this would have concerned everything from successful hunting to deciding where to seek shelter and how to outsmart predators. Monogamy also requires all of those social intelligence skills that we take for granted: the ability to listen and understand others, to lead or follow when necessary, to validate other people's feelings, to recognise that different people hold different beliefs. All of these demand a tremendous deal of energy that only humans and a few other brainy species can afford to expend. Only 9 per cent of mammals pair up for more than one breeding season, and only 15 per cent of primates live together as couples (and even they refuse to be sexually exclusive).

No one knows when monogamy began. The fossil record puts it somewhere between 3 and 4 million years ago, based on a few conspicuous clues. One clue was found in 1975, when the fossilised remains of seventeen members of the species *Australopithecus afarensis* were found in Hadar, Ethiopia. Thought to be around 3.5 million years old, the group comprised nine adults, three adolescents, and five children. They were found so close together they are believed to be part of the same family, and have since been dubbed 'the first family' (though this doesn't prove the parents were entirely faithful, of course).[21]

Another clue was unearthed in 2011, when archaeologists discovered that male and female finger lengths started to equalise around 3.5 million years ago.[22] This might not sound like much of a find, but finger-length equality is actually a telling sign of monogamous behaviour. In primates there is a connection between mating behaviour and body size differences between males and females of the same species. Biologists call it sexual dimorphism: the more dimorphic the primate, the more likely it will be polygamous (that is, having more than one mate). Male gorillas for example are twice as big as females and nearly always polygamous. Male and female gibbons on the other hand are equal in size and are mostly monogamous.

Some scientists think monogamy is even older, based on, of all things, penis shape. Humans are one of the few species to lack a penis bone, known scientifically as the baculum, which evolved around 90 million years ago and helps polygamous species have sex for longer, a crucial skill if an animal is to fertilise a female successfully amongst stiff competition (pun intended). In fact, the more straight, smooth and, frankly, boring a penis looks, the more likely the species is to be monogamous. The penises of polygamous species are usually far more exciting, with all kinds of twists, turns, kinks, ridges and spikes.

Homo sapiens are by no means a purely monogamous primate, however. According to recent studies, humans have evolved to be a monogamous primate that exhibits *tendencies* towards polygamy.[23] Scientists have devised a clever way of determining this. By using genetic data from three human populations – African, Asian and European – and looking at the frequency of X chromosome genetic material (which only recombines with females), they're able to measure the ratio of female to male breeding partners and how they have changed over time. A completely monogamous population would have a breeding ratio of 1:1 (one male to one female). The data churns out 1:1 in Asia, 1:3 in Europe and 1:4 in Africa, averaging about two women to every procreating man, which is still considered monogamous but nevertheless leans towards polygamy. Given that we descend from primates that are polygamous, this isn't surprising. Polygamy still echoes within the human genome and indeed most societies still practise some form of it. Modern monogamy has thus probably only existed for a few thousand years, certainly not long enough to silence the evolutionary impulse of polygamy completely.

How did monogamous brains evolve in the first place? Why were they so advantageous? Three schools of thought have emerged among scientists: female spacing, infanticide avoidance and a need for male paternal care.

Female spacing suggests that females deliberately spread themselves out across the savannah in order to retain a single male partner and obtain exclusive access to greater food resources. This is a win–win strategy for the pair because it means less fighting between competing males and ample security for the female's offspring since they are also protected from rival males. If our 4-million-year-old ancestor Ardipithecus, for example, had chosen to live an isolated existence, she might have fared better than mammals that reared their offspring in groups. In the brutal, unforgiving realm of the ancient savannah, her children would have been at high risk of being killed by other, more violent humans. By occupying a smaller territory that didn't overlap with other females, she could attract a male to guard her children; in exchange, the male secures a mate he can monopolise, ensuring paternity. This might sound harsh by the behavioural standards of twenty-first-century *Homo sapiens*, but it's important to remember that safety in numbers wasn't always a wise option for early humans, making female spacing an attractive alternative. As the Cambridge University zoologist Peter Brotherton observes in a paper supporting the female spacing idea, 'Provided that a single female can be monopolized successfully, monogamy can be viewed as a risk aversion strategy.'[24] But while this scenario is certainly possible, scholars admit that prehistoric females might have been too socially dependent on one another to spread themselves out in this fashion.

Infanticide avoidance posits that females needed monogamous partners solely to save their children from the aggressive and often murderous intentions of other males. Infanticide was an ever-present danger in the wild and is still practised by more than 40 per cent of primates: it's common among chimpanzees, gorillas and baboons; absent among orangutans, bonobos and mouse lemurs. Why it happens is unclear, but most scientists agree that it isn't blind violence. In primate society, infanticide might occur when food resources are too thinly spread, when a sick infant becomes a burden, or when the infant becomes food themselves and is cannibalised. A reliable male companion to defend a female's progeny was therefore essential. This too is a credible idea, especially since humans are also infanticidal primates. The ancient Romans, Greeks and Chinese killed both male and female babies in droves, as did medieval Europeans and Australian aboriginals. In fact, research conducted by anthropologist Laila Williamson reveals that infanticide has been practised on every continent and by every kind of people.[25] The reasons are similar to those of other primates: poverty (i.e. scarce resources), family planning (albeit a brutal form), perceived illegitimacy and/or the birth of infants with physical deformities. Still, the idea that infanticide pushes a species towards monogamy is probably false, since it remained the rule rather than the exception in monogamous societies; plus, many infanticidal primates have remained polygamous to this day.

Far more persuasive is the idea that monogamy evolved from the need for male paternal care. If a male redirects his energy from bellicose tribalism to pulling his weight at home, both he and his children are not only more likely to survive in the long run, but to develop into powerful individuals as well. Fundamental to paternal care is the notion of male guarding, that is, warding off rival males who may be desperate to find a reproductive-age female, especially if there is a shortage due to our male-biased sex ratio – there are roughly 105 males born for every 100 females (no one really knows why). If our ancestors experienced a scarcity of females, compounded by menopause and our long lifespan, male guarding was probably a crucial first step in the evolutionary pathway to paternal care, since it ensures paternity – which then provides a selection pressure for paternal care. In the brain, paternal care relies on the same structures that support maternal caregiving, namely the insula, the temporal and frontal cortex, the amygdala, the hypothalamus and the nucleus accumbens, which together form the 'human caregiving network'. This ancient network is rich in oxytocin receptors (receptors for one of our love hormones, discussed in the previous chapter), making it especially sensitive to caregiving behaviour.

For thousands of years, monogamous arrangements have been facilitated by the institution of marriage. While the rules and traditions of marriage have changed over time, one constant appears to be the pacifying effect it can have on males. For example, testosterone, a hormone known to promote male aggressiveness and poor decision-making, declines when males get married and have children. The sociologists Robert Sampson, John Laub and Christopher Wimer analysed data from a study that followed 1,000 low-income Boston teenagers for forty-five years and found that getting married made them 35 per cent less likely to commit crime (even after factoring in things that make marriage more likely: increased intelligence, financial stability, a healthy family background, etc.).[26] Today, no one would deny that males are just as able to raise children as females and gender non-binary people; indeed, 40 per cent of gay, lesbian, bisexual and transsexual people have children, and 60 per cent of those are biological. Splitting the responsibilities of childcare between two parents also lets the brain retain enough energy to pursue other activities, such as problem solving and creative thinking.

A touching example of paternal care is seen in the owl monkeys of South America. Male owl monkeys are involved in most of the childcare – carrying their infants on their back, feeding them and playing with them from the time they are two weeks old. A genetic study of seventeen pairs of owl monkeys and their thirty-five offspring revealed that every pair were their offspring's biological parents. They are completely monogamous, a trait that has been reported in only five species (the Malagasy giant rat, the California mouse, the oldfield mouse, the Kirk's dik-dik antelope and urban coyotes).[27] To preserve their emotional bond, the lifelong pair are often found with their long bushy tails draped around each other, snuggling in the treetop.

Dual parenting paves the way for others to pitch in. Anthropologists call them alloparents: unrelated group members as well as family members (think teachers, nannies, nursery staff, social workers and the kind strangers who help mothers with their pushchairs on the underground). No other species takes social thinking this far. But it makes perfect sense when we consider the big picture. With the help of alloparents human minds could get on with the business of building cities and civilisations. Alloparents are perhaps best encapsulated by the insightful African proverb, 'It takes a village to raise a child.' Popularised by the eponymous title of Hillary Clinton's 1996 book, it conveys the shared responsibility of parenting and the importance of community. It reminds us that no matter how small we feel as individuals, we all have a part to play. The Sukuma tribe of Tanzania have their own version: 'One knee does not bring up a child.' We only have to substitute 'knee' for 'brain' to unveil the metaphor. The fact is our social brains have evolved not just for our nearest and dearest but for the wider, extended family of humankind as well.

The philosophy of liberal parenting and the tendency for humans to act *in loco parentis* fashioned new kinds of social thinking. Like a great oak growing from a seed, human relationships became remarkably complex compared to those of other primates, and magnificent societies, from the bustling streets of Manhattan to the sprawling markets of Mumbai, burst forth. We may never know exactly how these forces shaped the evolving brain, nor the extent to which our brains have contributed and responded to the social advantages of monogamy. But new research shows that human monogamous brains are more active in reward-related areas (including the thalamus, nucleus accumbens, caudate, pallidum, putamen, insula and prefrontal cortex) than non-monogamous brains. Moreover, non-monogamous brains tend to have a special kind of dopamine receptor gene called *DRD4*, which is linked to promiscuity and infidelity. Of course, such findings face a chicken or egg conundrum since we still don't know which came first: the behaviour or the biology. As always, the answer is likely to be a complex interplay of both.

All told, human societies are remarkably complex compared to those of other animals. As we have seen, over the course of evolutionary history our social brains have equipped us with the perfect hardware for living in an increasingly crowded world. In our present age of globalisation, we can only hope that our minds are capable of extending and intensifying the social bonds that unite us. I for one am cautiously optimistic. The endless adaptability of the human brain should not be underestimated, and our increasing desire for social change is a powerful force.

'The arc of the moral universe is long,' proclaimed Martin Luther King Jr, 'but it bends towards justice.' That arc started with our 4-million-year-old ancestor Ardipithecus, and is limited only by the degree to which we are willing to cooperate with others. Yet as our brains evolved the traits of empathy, mutualism, compassion and fair play, this social complexity then became critically dependent on something else: memory.

4

The Genesis of Memory

Holly Shale prided herself on her memory: when she was fifteen, she won the silver medal in her junior memory championship for recalling a random sequence of thirty-five words. She had memorised all the capitals and flags of Africa. She remembered the atomic number and symbol of every element on the periodic table. She was a wonder child.

But when she was twenty-four, a car crash damaged her brain's hippocampus (a structure crucial for memory formation), producing anterograde amnesia. Unlike retrograde amnesia (the inability to recall past memories), anterograde amnesia is the inability to create new memories. Holly is thus perpetually locked in the present, powerless to continue the story of her life. For Holly, there are no tomorrows; there is only now and the feeling that something terrible has happened.

In March 2021 I video-called Holly at her home in east London. Her mother, Elizabeth, took the call in the kitchen. Holly was hunched over the counter in a blue apron, whisking eggs and staring at a cookery book. She was tall and wiry, with almond eyes and dark frizzy hair. Elizabeth introduced us and then said to Holly, 'I think this will be the one, darling. What number is this?'

'I'm not sure,' she said. 'I keep getting the mixture wrong.'

'Ah, right. Well come sit down and we'll make the cake later.'

There were four cakes on the kitchen table; according to Elizabeth, she will bake a cake, serve it up and head straight back to the counter to make another one. Elizabeth accompanies her when she leaves the house to stop her buying the same thing again and again. She is forty-two, but still thinks she is twenty-four.

'I'm a bit like Dori from *Finding Nemo* now,' she laughs. 'It's not too bad, as long as I've got someone with me. In the beginning it was awful. Every day was a blank slate. I would read a newspaper article and instantly forget what it was about. I would wash my hair over and over again. I just couldn't function. The only reason I'm able to tell people any of this is because of the care and support of my mum.'

As I listened to Holly's story, noting how her mother often guided her towards what she was trying to say, I was struck by how resilient and courageous she had become. Here was a woman whose mind was caught in an abbreviated version of life, a woman who now relied on others to be her memory. I was reminded of my grandfather, whose battle with Alzheimer's disease obliterated his memory in the reverse direction, and how, for a time, he too was forced to remember life vicariously through his family. But what struck me most about Holly's plight was that it shattered all our preconceptions about memory.

The idea that memory is about the past – a giant filing cabinet in which the files are past experiences, ready to be accessed with the right cues – stretches all the way back to ancient Greece. Plato viewed memory as a wax tablet: our memories and thoughts could be stamped, stored and retrieved for later use. Socrates likened memories to birds in an aviary, 'some in exclusive flocks, some in small groups, and some flying alone, here, there and everywhere among all the rest.'[1]

Surprisingly, it turns out that memory has almost nothing to do with the past. Memory is about the present and the future. From the moment you are born you are constantly encountering new phenomena in the world, which your brain interprets and organises into experiences. They include our physical surroundings, our bodily sensations, the norms and values of social life, and the ever-changing stream of subjective feelings that only ceases at death.

When we reflect on past experiences, we are not merely taking a trip down memory lane, we are also accessing an earlier brain state in order to influence our current or future behaviour. The act of remembering something is a type of learning; it's a constructive process, allowing us to predict the future based on memories of the past. This is why we are such poor witnesses in court: our future-orientated memory is better at imagining possible scenarios than remembering real ones.

It's often said that memory makes us who we are; but evolution doesn't care about the personal history of each human brain. It cares only about the changes that have occurred throughout the brain's 7-million-year journey, and whether those changes give us a better chance to survive and reproduce. Consequently, the story of human memory contains a number of important subplots – each concerned with a monumental shift in memory.

The Seven Stages of Memory

Every animal has some kind of memory. Elephants, for example, can identify at least thirty of their relatives and are known to remember the Maasai tribe members who abuse them in ritual displays of masculinity. Crows including ravens and magpies recognise individual human faces, ignoring those considered friends and scolding those deemed foes. Single-celled organisms such as the paramecium, a water-loving animal covered in microscopic hairs, will learn to avoid being electrically shocked in lab experiments, encoding the memory in a web of organic chemicals.

These observations tell us that memory is old. Very old. Most researchers believe that human memory was born in the ancient oceans. Free-floating neurons coalesced to form so-called nerve nets, seen today in jellyfish, sea anemones and corals. Scientists are very interested in these creatures. If we can decipher the code by which their neurons generate memory, we may one day be able to place that capacity onto a silicon chip and implant it into patients with memory disorders. Memory 'neuroprosthetics' might sound far-fetched, but governments in the Western world are investing tens of millions of dollars into it (of which more anon).

It used to be thought that all kinds of memory – recalling your name, where you live, what you ate for dinner last night, how to swing a tennis racket – were all the same mechanism in the brain. Memory was regarded as a simple phenomenon concerned only with the automatic and unconscious storage of information. Now, however, we are learning that memory has experienced an evolutionary path as long and winding as every other faculty of the mind. Scientists believe that seven different memory systems evolved in order to help the brain solve problems and exploit opportunities encountered by early humans in the distant past. These memory systems are thought to have started their journey in early animals, but then evolved independently in humans some 3 million years ago, when four species of human ancestor – *Australopithecus africanus*, *Australopithecus bahrelghazali*, *Australopithecus deyiremeda* and *Kenyanthropus platyops* – roamed the earth.

The first system of memory to evolve was reinforcement memory: remembering predators and the location of food sources.[2] In the deadly proving grounds of the savannah, learning what – and who – our enemies were and how to feed our children was critical. Modern examples can be seen in the behaviour of any prey–predator relationship you care to name: zebras must remember to avoid lions, deer to avoid wolves, rabbits to avoid foxes, and so on. A long history of bloodshed slowly taught animals which actions paid off and which came at a price. The biology behind this type of learning was a mystery until Ivan Pavlov (1849–1936) famously showed that dogs could learn to associate a particular stimulus (a bell) with another event that followed (being fed), causing them to salivate. Pavlov called this response psychic secretion, but psychologists now know it as reinforcement memory. The neurobiology behind it has always been a puzzle. We know that the cerebellum (the 'little brain' located at the back of the brain, just behind the brain stem) must be involved, given that remembering movements – for example, the movements needed for escape – is clearly important. And we know that other brain regions such as the amygdala are involved, given that fear and other emotions also play a role. But that's it. Like much in neuroscience, we know a lot about what and where but much less about why and how.

Closely related is the second system – that of navigation memory: our ability to build a cognitive map of the world. While Google Earth is certainly impressive, its processing power pales in comparison to the millions of place cells in the hippocampus, each firing in response to the smallest change in our environment – a pointillism of sights, smells, objects and places. They belong to a class of neurons called pyramidal neurons, which become active when we enter a specific location, collectively mapping the location in a pattern of neuronal activity. We take this memory for granted – remembering the journey to work or where we parked the car rarely feels impressive – but life has gone to extraordinary measures to invent it. Migratory birds for instance use a magnetic compass in their eye for navigation memory. Honeybees use a map of polarised light, visible even on the cloudiest of days. Bats and many other mammals use echolocation. In humans, place cells allowed our ancestors to carve out new territory and become world travellers, journeying all the way from Africa to Asia.

Next came something called biased-competition memory: the ability of various memories to compete for our attention. The brain pulls off this trick by encoding memories in neurons that suppress the activity of neighbouring neurons. In fact, memories formed within six hours of each other activate the same group of neurons, which then try to suppress neurons encoding memories formed in the more distant past. In this way, our brains can select which memories from our past are worth keeping. When early humans had to decide between various hunting strategies, weighing up the pros and cons of each based on previous experience served them well. More than 2 million years ago, *Homo habilis* used this faculty to ambush herds of antelope and gazelles. Experts now believe that they hid in trees and leapt on their prey from above, spearing the animals at point-blank range.[3] For this method to prevail, memories of successful hunts would have had to outcompete memories of failed hunts.

Competition memory doesn't mean that we lose out-of-date memories, only that we can build on them. By learning which situations required an up-to-date approach and then creating a bias towards that approach, human memory developed a level of ingenuity not seen in other primates. So can we choose what we forget? Yes, according to Jeremy Manning at Dartmouth College, New Hampshire. He's shown that simply telling people to 'push thoughts out of their head' is enough to make them forget a list of words, and he is now helping patients with PTSD (post-traumatic stress disorder) to forget painful memories that keep resurfacing. The neuroscience underlying this process is unclear, but it probably takes place at the synapse. When memories are formed, synapses strengthen their connections by increasing the number of a particular type of receptor called AMPA. These receptors are notoriously unstable, constantly shuttling in and out of the synapse. To forget something, our brains must actively remove these receptors, weakening and eventually destroying the connections at the synapse. Manning observes: 'Forgetting is typically viewed as a failure in some sense, but sometimes forgetting can be beneficial, too . . . we might want to get old information out of our head, so we can focus on learning new material.'[4]

Fourth to evolve was a system for remembering how to grasp certain objects in the thin branches of the trees, called manual foraging memory, or muscle memory. This would later become an important way of creating objects such as stone tools and fishhooks. While reaching for your morning coffee certainly doesn't feel dependent on memory, experiments on a patient named

Henry Molaison (known to psychologists as HM) have proved otherwise. In 1953 HM had two-thirds of his hippocampus surgically removed to treat his epilepsy. The procedure worked but left him in the same predicament as Holly Shale: unable to form new memories. In 1998 the neuroscientist Reza Shadmehr and his colleagues trained HM to grasp a robotic handle that was designed to resist being handled. After the experiment, HM quickly forgot who the experimenters were and claimed never to have seen the robotic device before. And yet, when the team asked him to try grasping the robotic handle again, he knew exactly how to perform the task without any help.[5] In the brain, this memory relies heavily on the cerebellum; specifically, a special class of neurons within the cerebellum called the Purkinje cell, named after the Czech anatomist Jan Evangelista Purkyně who discovered them in 1839. Purkinje cells are instantly recognisable due to their vast, intricately branched network of dendrites, which creates muscle memory by constantly remodelling itself in response to neurons projecting from the spinal cord. Some scientists believe this memory evolved to let us quickly rebuild muscle mass during times of strife, freeing us from having to maintain a constant warrior physique.

Once we had evolved memory based on touch, evolution turned to sight and hearing. For this we evolved feature memory, a sensory memory for important visual and auditory cues. For example, the sight of a cave usually signified sanctuary, the growl of the sabre-toothed tiger peril. For this system to work, human brains had to interpret sight and sound at the same time, even though they are very different senses. The brain does this by merging electrical activity from the visual and auditory cortex, combining it in a special population of neurons called polychronous neuronal groups (PNG). These neurons 'polychronise', that is, produce a regular and repeating firing pattern, which we think lets the brain combine different sensory inputs into a single experience. Sometimes this merging goes too far and generates synesthetes: people who taste words, hear colours and see sounds. Synesthetes have an enhanced memory compared with the rest of the population. Some believe their richer perceptual experience improves their feature memory.

As human species became more diverse, between 500,000 to 2 million years ago, their territory expanded to occupy almost all of the planet's habitable regions. Waves of human migrants, including *Homo erectus, Homo heidelbergensis*, Neanderthals and *Homo sapiens*, left Africa, swept across the Red Sea to the Middle East and Australia, up into Europe and across the Bering Strait into the Americas. Along the way *Homo sapiens* developed large brains, we lived longer and we evolved social thinking to form tribes. And for all this we evolved goal memory: the power to remember certain objectives and targets of action, a memory to allow our thoughts to create the future. Research points to the lateral prefrontal cortex for this type of memory: neurons in this region fire in response to reward and decision-making, allowing our brains to code the value of different goals and decisions for the future. Its applications were boundless. If a fellow *Homo sapiens* remembered to plant seeds instead of eating them, the whole tribe would benefit. If a fellow *Homo sapiens* remembered which rivers had the most fish, the knowledge could be passed on. And if a fellow *Homo sapiens* laid claim to more than their fair share of resources, that was worth remembering in order to prevent it occurring again.

Which brings us to the last system to evolve: the memory of one's behaviour and its consequences in social systems, known as social subjective memory – another ingredient vital to the evolution of social minds, partly forming the basis of human morality. This kind of memory is thought to have evolved only a few hundred thousand years ago, as a means of adapting in early *Homo sapiens* societies. We all have to curb and sometimes alter our behaviour to make group living more harmonious. And those constraints depend on memories of good or bad outcomes in social situations. Let's say you are hosting dinner for new neighbours and make a joke about accountants. They laugh politely, but you later discover during the course of the dinner that one of your guests is, in fact, an accountant and was offended by the joke. If you are a good neighbour, chances are you won't make a similar joke in the accountant's company again, not out of shame but because you remember the negative consequences of your behaviour. This is what psychologist Daniel Kahneman calls the 'remembering self'. It's how our brains plan for the future. The next dinner won't be awkward. You know because you can already see it. As Kahneman notes, 'We think of our future as anticipated memories.'[6] The brain does this using a region of the parietal lobe called the precuneus, though the underlying neurobiology of this memory remains unclear.

To see how all these systems work together today, let's imagine a typical day in the life of a modern human. Every day you wake up in the morning, get ready for work and walk out of the front door. What confronts you is not a savannah or wilderness; there are no wild dogs or big cats waiting to have you for breakfast, but typically a road with some cars gliding past. Here is where your reinforcement memory kicks in. Car accident deaths are common and deadly, making vehicles the most immediate 'predators' preventing you from reaching your 'food source' or job income.

After safely crossing or avoiding the road, you now need your navigation memory to direct you to work. Previous journeys have of course hardwired the route in your brain, but you still need your hippocampus's place cells to light up in just the right sequence to get you there. If you walk or take public transport, you might decide to use the time on your way to work browsing on your phone or reading a book. The various options competing for your attention are your biased-competition memory speaking up. Perhaps you have an email that is overdue for a reply, in which case you'd better put the book away.

Everything you've been doing thus far on your journey depends on muscle memory, or manual foraging memory, without which you wouldn't even have remembered how to brush your teeth or put on clothes. And if you're planning a trip to the gym after work, your capacity for this type of memory will determine how successful your session is.

By now, you've arrived at work, an office filled with the sound of people talking, telephones ringing, pages turning, computer keys tapping and electric fans whirring. Add to this the sight of screens flickering, colleagues mingling, deliveries arriving and meetings in full flow, and were it not for your feature memory, meticulously categorising every auditory and visual detail, you might almost succumb to sensory overload.

Settled in at the office, you can now get on with the business of actually doing some work. No doubt you have a long list of objectives and targets to meet, all of which will demand your goal memory: your ability to use your thoughts to create the future. The key difference between you and one of your ancestors in this scenario is that your ancestors were hunting game, not clients.

All that's left for your memory to worry about now is how you're coming across to your colleagues and clients. Like a referee at a football match, your social subjective memory is keeping a close eye on your behaviour and its likely consequences. A flood of anticipated memories fills your mind, showing you all the ways in which your day could be a triumph or a shambles.

Memory gets us through each day in more ways than we know, and these systems are probably only the tip of the iceberg. Overall, the history of memory in humans is closely tied to the history that made us the complex organisms we are. There are 37 trillion cells in the human body, controlling everything from moving to thinking. Memory developed through stages of evolution to complement these functions and help humans react appropriately to every situation they encountered. A brain that can learn and memorise and make decisions based on prior experience is a quantum leap in biology. It's what makes our species responsive to the environment rather than reflexive, purposeful rather than aimless. Memory is a story that has been endlessly edited, updated and retold for a new audience.

How Memory Works

Throughout history, three important breakthroughs shaped the course of memory research. In 1749 David Hartley proposed that memory was a product of the nervous system and other biological processes. In 1904 Richard Semon proposed that experience leaves a physical trace in the brain, but that these memory traces were imperfect copies of experience, prone to distortion. And in 1932 Frederick Bartlett proposed that what people remember is based on our cultural values and the way we perceive the world around us: white Americans, for example, tend to stress the role of individuals in their descriptions of memories, while East Asians focus more on social interactions. There were more advances in the years following – in 1949 Donald Hebb showed that durable connections between neurons could encode memory and, in 1974, Eric Kandel showed that the substance of memory is protein molecules and neurotransmitters obeying the principles of biochemistry. But despite all this progress, the biology of memory remained unclear.

For simplicity, we now categorise memory into several different types. Working (short-term) memory is fairly self-explanatory. It refers to all the information you temporarily hold in your head, for instance remembering numbers when you do a sum or holding a person's address in your mind. It lasts around ten to fifteen seconds and can only store a limited amount of information: most people, for example, can only accurately hold seven numbers in working memory at any given time. It relies on the frontal lobes in particular but also appears to involve most of the cortex; the neurophysiology is poorly understood. A useful analogy is to think of working memory as RAM in a computer, whereas long-term memory would be the hard drive. Not everyone's working memory is the same. One in ten children has working memory deficits, meaning they perform poorly in school despite being intelligent otherwise.[7] And though not proven, many experts believe working memory deficits are linked to ADHD (attention deficit hyperactivity disorder).[8]

Then there's long-term memory, which can be divided broadly into explicit (or declarative) and implicit (or procedural). Explicit memory refers to factual information about your life – your parents' names, where you grew up, what you do for a living – and can be consciously recalled (or 'declared'). It can be further divided into episodic memory, the memory of past experiences and events, and semantic memory, the memory of knowledge gathered over a lifetime. Implicit memory is the memory of how to perform certain tasks, such as walk, talk, ride a bike or play the piano. Starting from very early in life, implicit memory can become so ingrained it is almost automatic. No one knows how the brain stores long-term memory, but it is almost certainly not in one place. Some of the most common brain diseases such as Alzheimer's and stroke obliterate long-term memory by destroying vast swathes of brain tissue. It seems that long-term memory behaves more like a Wi-Fi network than a hard drive.

Many scientists argue that memories are physically encoded in the brain by a network of neurons in the hippocampus and cortex. The fundamental idea is that you have an experience – say, you lose your virginity – and a memory of the experience is sent to the hippocampus in the form of electrical signals travelling across synapses. The memory may then reside in the cortex as long-term memory, or it may not. That depends on complex molecular processes involving neurotransmitter receptors, enzymes, genes, epigenetics, and so on. Until recently this idea was mainstream neuroscience. Pursuing it was thought to be our best hope of finally understanding how memory works.

But I and many other neuroscientists are now convinced that it is wrong, thanks to a brilliant investigation by the neuroscientists Nikolay Kukushkin and Thomas Carew in the journal *Neuron*.[9] Kukushkin and Carew were eager to challenge the mainstream theory when they noticed a fallacy in the way we talk about memory. The language that scientists use to describe memory – 'memory retrieval', 'memory acquisition', 'memory trace', 'memory consolidation' and so on – presupposes the notion of 'a memory', something that is separate from the person doing the remembering. Do you see the problem? The language defines memory as separate from the mind instead of an integral part of it; when the truth is, your brain doesn't store or retrieve memories. It *is* memories.

If this has left you confused, or so bewildered that you feel we are tumbling down the rabbit hole, you are not alone. The world's leading scientists don't understand it either. The best explanation is that memory must be a change from one brain state to another. And so the act of remembering, Kukushkin notes, is 'just a reactivation of connections between different parts of your brain that were active at some previous time.' Neuroscience now accepts this idea, but it is astonishing to most people when they first ponder it. One almost feels like Proust as he bites into the famous madeleine, reviving his childhood memories and distorting the boundaries of time:

> But at the very instant when the mouthful of tea mixed with cake-crumbs touched my palate, I quivered, attentive to the extraordinary thing that was happening in me. A delicious pleasure had invaded me, isolated me, without my having any notion as to its cause . . . Where could it have come to me from – this powerful joy? I sensed that it was connected to the taste of the tea and the cake, but that it went infinitely far beyond it . . . And suddenly the memory appeared. That taste was the taste of the little piece of madeleine which on Sunday mornings at Combray . . . my aunt Léonie would give me after dipping it in her infusion of tea or lime-blossom.[10]

Proust's bewitching epiphany reveals another puzzling feature of memory. It suggests that neurons, synapses and molecules can sense the passing of time. Though we might recognise this phenomenon from our body's circadian rhythm that is its own kind of time-keeper – an automatic process designed to regulate sleep and wake-fulness – our new understanding of memory, on the other hand, reveals that the brain resurrects our conscious experience of past realities whenever it wants.

This suggests that the brain also invented memory to measure time. Indeed, we now know that the hippocampus contains what researchers call 'time cells': neurons that represent particular moments in time as well as the location of specific experiences. If the hippocampus is damaged, as it is in patients with Alzheimer's and other memory disorders, people can only accurately recall the passage of time for short intervals. This is why Alzheimer's destroys short-term memory first. Yet the question remains: how does the brain measure long time periods?

This is where things get a little complicated. It turns out that the brain contains multiple inner clocks, some for tracking very different time periods (1 or 10 minutes), some for tracking very similar time periods (10 or 12 minutes), some for tracking short time periods (5–10 seconds), and some for tracking longer time periods (20–60 minutes). And all of these clocks compete for our attention.

To see how the brain discriminates between them, a University of California neuroscientist named Norbert Fortin and his colleagues trained rats to signal time.[11] To do this, they taught the rats to choose between different odours: odour A for a one-minute interval, odour C for a twelve-minute interval and odour B for the intermediate interval. As an incentive, tasty treats were given to the rats that got it right.

Before the experiment, the scientists administered a drug that briefly shuts down the hippocampus. This kind of interference tells them whether the hippocampus is required to perform the task accurately. They found that the hippocampus is necessary for discriminating between similar longer periods of time (say, 20–24 minutes) but not for discriminating between events occurring second by second. This is evidence, according to the authors of the study, that the hippocampus contains an inner clock for separating individual experiences during a sequence of events. So when, for example, you finish cooking supper and then sit with the family at the dinner table, your hippocampus is in charge. But when sensing the difference between cooking a steak that's rare versus medium-rare, something else in the brain is active.

That something is thought to be the striatum, a region found deep within the front part of the brain. Most scientists believe that its neurons evolved to code time with extreme precision. Without it, we would be lumbering through everyday tasks like a paralytic drunk. Just imagine trying to write an email without remembering the first half of it. Imagine tying your shoelaces without remembering why you put your shoes on. Life would be almost impossible. For all its stages of evolution, memory is useless without a strong grip on time. This fact was vividly portrayed in the 2000 Christopher Nolan thriller *Memento*. The protagonist, an insurance broker named Leonard Shelby, is hit on the head by an intruder and soon develops anterograde amnesia. He then becomes obsessed with catching the attacker, who he believes raped and murdered his wife, but because his memory is wiped clean every fifteen minutes, he must have every clue to the murder painstakingly tattooed to his body. At one point in the film, he perfectly captures the essence of his distress: 'How am I supposed to feel, when I can't feel time?'

Cultural Memory

One of the most astonishing things about memory is how much it is shaped by culture. One study found that white Americans are more prone to false memories than people from Eastern cultures.[12] The pioneer of this field, Frederick Bartlett, demonstrated the importance of culture by asking English subjects to memorise and then recount a North American folktale called 'The War of the Ghosts':

> One night two young men from Egulac went down to the river to hunt seals, and while they were there it became foggy and calm. Then they heard war cries, and they thought: 'Maybe this is a war-party.' They escaped to the shore, and hid behind a log. Now canoes came up, and they heard the noise of paddles, and saw one canoe coming up to them. There were five men in the canoe, and they said:
>
> 'What do you think? We wish to take you along. We are going up the river to make war on the people.' One of the young men said: 'I have no arrows.' 'Arrows are in the canoe,' they said. 'I will not go along. I might be killed. My relatives do not know where I have gone. But you,' he said, turning to the other, 'may go with them.' So one of the young men went, but the other returned home.
>
> And the warriors went on up the river to a town on the other side of Kalama. The people came down to the water, and they began to fight, and many were killed. But presently the young man heard one of the warriors say: 'Quick, let us go home: that Indian has been hit.' Now he thought: 'Oh, they are ghosts.' He did not feel sick, but they said he had been shot.
>
> So the canoes went back to Egulac, and the young man went ashore to his house, and made a fire. And he told everybody and said: 'Behold I accompanied the ghosts, and we went to fight. Many of our fellows were killed, and many of those who attacked us were killed. They said I was hit, and I did not feel sick.'

> He told it all, and then he became quiet. When the sun rose he fell down. Something black came out of his mouth. His face became contorted. The people jumped up and cried. He was dead.[13]

Bartlett knew that this story would appear disordered and somewhat inscrutable to Anglo-Saxon minds. Every culture, he argued, has what he called a schema: a framework of thought that helps organise and interpret knowledge. If we are unsure about what has happened, we use our own schema to fill in the gaps and rationalise what we are trying to remember. And indeed when Bartlett asked his subjects to reproduce the story, he found that many features had been added, subtracted, simplified and transformed. Here is one attempt:

> Two youths went down to the river to hunt for seals. They were hiding behind a rock when a boat with some warriors in it came up to them. The warriors, however, said they were friends, and invited them to help them to fight an enemy over the river. The elder one said he could not go because his relations would be so anxious if he did not return home. So the younger one went with the warriors in the boat. In the evening he returned and told his friends that he had been fighting in a great battle, and that many were slain on both sides. After lighting a fire he retired to sleep. In the morning, when the sun rose, he fell ill, and his neighbours came to see him. He had told them that he had been wounded in the battle but had felt no pain then. But soon he became worse. He writhed and shrieked and fell to the ground dead. Something black came out of his mouth. The neighbours said he must have been at war with the ghosts.

This may not strike you as very surprising, because we all know that memories can be reshaped like a game of whispers. Memories change each time we remember, often more than we realise. In fact, research shows that when we tell stories we actually change little details depending on the listener's personality and political outlook, a phenomenon known as the audience-tuning effect. This then later changes how we ourselves remember the story, further spinning the wheels of disinformation. To counter this, most people try to rehearse a memory immediately after an event. But even this act makes us more susceptible to misinformation later, a phenomenon known as retrieval-enhanced suggestibility, which is a real nuisance to police officers who later discover that an eyewitness generated details that were completely false. The disappointing truth is that human memory is only ever as reliable as the most recent story we tell ourselves.

Bartlett's discovery that human recollections are coloured by social constructs including race, language, education, and the lived experience of the individual is one of the most exciting ideas in neuropsychology. It proves that none of us is in a privileged position to define what is true. Our memories, like our emotional and social minds, are bound together as yet another adaptation of our culturally evolving minds. This is not to say that 'what really happened' does not exist; it does. The bedrock of good science is, after all, objectivity. The fact that different cultures remember the world around them differently says more about our social behaviour than our memory. It also says that for *Homo sapiens* the purpose of memory is partly to serve the collective.

Collective Memory

There is a kind of memory that has been with us since the early days of our species; the kind stored in leather scrolls and wax tablets, books and art, computers and smart phones. We call it collective, or social, memory. It arises when individual memories are shared by members of a community, and when a community then enshrines those memories in public symbols and institutions.

The first to study it seriously was the French philosopher Maurice Halbwachs (1877–1945), who argued that memory only works within a collectivist context – that memory paints images in the mind only 'in accord with the predominant thoughts of society.' Unlike the types of memory that evolution took eons to devise, collective memory is constantly changing depending on the historical narrative of our species. As a group, we remember all kinds of events: the Holocaust, the fall of the Berlin Wall, the moon landing, the end of apartheid, 9/11. This is because members of a group – from small groups such as families to large groups such as nations – usually share similar memories. Americans, for example, remember their country's dark history of slavery; Shia Muslims remember the death of Ali (regarded as the first imam after the prophet Mohammed); and my family remember my grandfather's long struggle with Alzheimer's disease.

As social animals, we are continually adjusting our memory of the past to conform to social precepts. Memory is a social tool. When humans share memories with one another they are creating a shared reality, one that is concerned less with the fidelity of the memory and more with whether or not the speaker and listener belong to the same social group. Like individual memory, collective memory is represented by neurons and synapses in the brain. In an attempt to discover the underlying neurobiology of this process, Micah Edelson, at the Weizmann Institute of Science in Rehovot, Israel, and his colleagues asked thirty people to watch an eyewitness-style documentary about a police arrest in groups of five.[14] A few days later, the subjects returned to the lab and completed a memory test about the documentary while being scanned with an fMRI machine. At this point, though, the researchers got creative. The subjects were offered a 'lifeline': the answers given by their fellow observers. That's what they thought at least, because the answers were actually composed of false memories.

Astonishingly, nearly 70 per cent of the subjects changed their answers to fit in with the group. The question is, did they *knowingly* change their answer to conform? Earlier studies of this kind have shown that people do indeed bow to social pressure and give a false answer, even though they privately hold a different belief.[15] So the researchers asked the subjects to take the memory test again, only this time they told them that the previous answers given by their co-observers were in fact randomly generated by a computer. No social pressure, so no need to lie. The result: some of the subjects admitted what they really remembered, but 40 per cent remained convinced by the false memories. They still felt completely confident that their memory was accurate.

When Edelson and his team looked at which brain areas might be active during such puzzling behaviour, they found a strong co-activation between the hippocampus and the amygdala. This suggests that social pressure can actually change your memories. The endless desire to fit in, these scientists think, may even be the brain's stamp of approval for a memory to be formed at all. As to why our brains evolved this way, they conclude:

> Memory conformity may also serve an adaptive purpose, because social learning is often more efficient and accurate than individual learning. For this reason, humans may be predisposed to trust the judgement of the group, even when it stands in opposition to their own original beliefs. Such influences and their long-term effects . . . may contribute to the extraordinary levels of persistent conformity seen in authoritarian cults and societies.

Perhaps unsurprisingly, collective memory is coloured more by a person's lived experience of an event than by later accounts in the history books. Lived memories describe events in personal terms and are often imbued with strong emotion and meaning. Non-lived memories, on the other hand, are usually described in abstract, matter-of-fact terms. Psychologists interested in this phenomenon study what they call temporal construal theory, which argues that the more psychologically distant one is from a historical event

– that is, the more non-lived their memories are – the less likely they are to appreciate fully what really happened. This matters because it affects how we remember history. Someone may know a great deal about the communist horrors of the Soviet Union, for example, but this is a poor substitute for someone who actually lived through them. Equally, some young Germans have been found to excuse their grandfather's Nazi affiliations because the fascist horrors of Nazi Germany and the murder of 6 million Jews is – for them – an abstraction rather than a lived experience.[16]

Unsettling though it may be, the truth is that memory is not what we think it is. It twists and distorts our perception of reality in countless ways. And if our brain's tribal nature is not ultimately overcome, it may continue to deceive us for a very long time.

As we have seen, memory was born in stages of evolutionary development to help our ancestors respond to an ever-changing material and social world. Along the way, our brains understood that the memory of past experiences could be put into the service of the present and the future, ensuring reproductive success by anticipating the sequence of events that a human life was likely to contain. Our minds have trouble understanding memory, so most people falsely identify it as something living within us – something contained, retrieved and then filed away in the brain. But in reality, we *are* our memories. The brain is as much memory as the skeleton is bone. As the novelist John Irving has put it, 'You think you have memory; but it has you!'

Our understanding of memory is also enriched by the discovery that culture shapes how we remember our past. If your culture values social interactions over individualism, for example, your memories will be influenced accordingly. In the same way, the study of memory has been enriched by the discovery that brains generate collective memories to create a shared reality for the group, which has important consequences for how we remember our past and shape our future.

What followed memory in the brain's evolution was an event that set us apart from all other primates. For reasons that we are only just discovering, the modern human brain underwent a staggering and rapid transformation in cognitive and behavioural abilities some 200,000 years ago. Described by scientists as the Human Revolution, it would prove to be the wellspring of our species' intellect, language and creativity.

5

The Truth About Intelligence

In 1905 seventeen-year-old Indian mathematician Srinivasa Ramanujan ran away from home. Failing almost every subject other than maths, he lost his scholarship at the prestigious Government Arts College in Kumbakonam, a town in Tamil Nadu, India. His interest in mathematics had become an obsession, and he spent nearly all his free time solving theorems and discovering new ones intuitively. Two years later, he gave up on formal education entirely, pursuing maths in private while living on the verge of starvation.

In 1909, hoping to gain some kind of employment, Ramanujan visited the Indian Mathematical Society. Its president, a man named Dewan Bahadur R. Ramachandra Rao, described the encounter:

> A short uncouth figure, stout, unshaved, not overclean, with one conspicuous feature – shining eyes – walked in with a frayed notebook under his arm . . . [He] showed me some of his simpler results. These transcended existing books . . . he led me to elliptic integrals and hypergeometric series and at last his theory of divergent series not yet announced to the world converted me. I asked him what he wanted. He said he wanted a pittance to live on so that he might pursue his researches.[1]

Ramanujan dazzled and baffled his peers. Exhilarated, they suggested that he send his work to Professor Godfrey Harold Hardy of Trinity College, Cambridge, himself an eccentric maths prodigy who, as a child, would write numbers up to millions and amuse himself in church by factorising the hymn numbers. When Hardy opened Ramanujan's letter, one morning in early 1913, the theorems he saw appeared wild and unthinkable, surely the musings of a crank. But it didn't take long for Hardy to realise the truth: Ramanujan was a man of untrammelled genius. His formulas 'defeated me completely', Hardy later admitted. 'I had never seen anything in the least like this before . . . they could only be written down by a mathematician of the highest class.'[2]

Hardy and Ramanujan formed a legendary partnership, transforming twentieth-century mathematics and contributing to fields virtually unheard of in their lifetime, including quantum computing and black hole research. More than anything, their story demonstrates that of all the changes the human brain has experienced, its increased intelligence is perhaps the most spellbinding.

When we think about how humans became intelligent, we often picture small bands of hunter-gatherers no brighter than Homer Simpson gradually transitioning to a lifestyle of farming, settlement and industry. But the agricultural and industrial revolutions occurred long after the brain changes that made them possible.

In order truly to understand intelligence we must go back to its origins. Specifically, we must study intelligence in its proper context: that of a shifting African climate filled with uncertainty and hardship. When the African climate see-sawed between wet and dry periods, as it did 4 million years ago, ancestors such as *Australopithecus* needed the wits to adapt to their immediate surroundings. When the African landscape shifted towards open savannah, as it did 2 million years ago, ancestors such as *Homo ergaster* needed the shrewdness to make sophisticated stone tools and plan long-distance hunts. And when the African climate became too dry for human habitation, as it did 200,000 years ago, *Homo sapiens* needed the foresight temporarily to leave Africa and spread across the world.

The earliest evidence of human intelligence dates back roughly 2.5 million years, when *Homo habilis* used what are now considered the oldest human inventions: a set of stone tools discovered at Olduvai Gorge in Tanzania. Known simply as the Oldowan, they are rough, all-purpose tools, pointed at one end and were fashioned by smashing two rocks together to release sharp flakes of stone. Archaeologists call them choppers. The Oldowan were probably used by *Homo habilis* to split nuts and fruit and butcher small animals. The fact that they were all-purpose tools suggests that *Homo habilis* were not particularly intelligent. A sophisticated stone tool is one built with a particular task in mind, such as a hammer or scraper.

In all likelihood *Homo habilis* had an intellect that cognitive scientists call pre-representational, meaning they were clever enough to invent tools based on their sensory experience of the world but not clever enough to engage in derivative or abstract thinking. Add to that a 600-cm^3 brain size (less than half that of *Homo sapiens*) and an underdeveloped memory, and things start to look rather bleak for *Homo habilis*. As evolutionary psychologists Liane Gabora and Anne Russon write,

> They could store perceptions of events and recall them in the presence of a reminder or cue, but they had little voluntary access to episodic memories without environmental cues. They were therefore unable to voluntarily shape, modify, or practise skills and actions, and they were unable to invent or refine complex gestures or means of communicating.[3]

Human intelligence had its first real growth spurt when *Homo erectus* appeared, some 1.8 million years ago. The brains of *Homo erectus* were 25 per cent larger than that of *Homo habilis* and 75 per cent the size of *Homo sapiens*. In consequence, a variety of distinctly human behaviours emerged. They made sophisticated stone hand-axes including the Aschulean hand-axe, a symmetrical blade that required several stages of production. They were the first to use fire and possibly the first to cook. They may even have been the first to use rafts or other seagoing vessels to leave Africa for Europe and Asia. All of which is far more significant than the cognitive achievements of earlier hominins (or indeed of other primates), for they involve the ability to reason, plan and grasp deep truths about the world.

It's also thought that *Homo erectus* possessed brain regions for language. While such language was probably very simple (just a few words with no linguistic structure), it does indicate a mode of thinking unprecedented in the hominin lineage. Scholars call this mode mimetic cognition (Greek for mimesis, 'to imitate'). The brain's capacity for mimetic cognition meant that *Homo erectus* were no longer trapped by their immediate sensory experience, as *Homo habilis* had been. Instead, they could act out events that happened in the past, rehearsing and learning and planning for the future in a truly predictive fashion. Gabora and Russon explain:

> [Mimetic cognition] enabled hominins to engage in a *stream of thought*. One thought or idea evokes another, revised version of it, which evokes yet another, and so forth recursively. In this way, attention is directed away from the external world towards one internal model of it.

Then something remarkable happened. Somewhere between 200,000 and 30,000 years ago, intelligence skyrocketed with the ascendancy of *Homo sapiens.* Described by anthropologists as the Human Revolution, this period saw more innovation than the previous 7 million years of human evolution. Our species developed language, music, religion, art and trade. We made the seemingly impossible journey to Australia, and combined our collective intelligence constantly to improve our technology in a process that scholars call the ratchet effect: the snowballing of progress that occurs when good ideas jump from one mind to another, until the whole population is ratcheted up. What's more, we invented perhaps the defining feature of human existence: metacognition – the power to think about thinking.

How do we explain the surge of intelligence that allowed *Homo sapiens* to conquer the world? Neuroimaging offers a clue. The cerebral cortex is so large that it had to wrinkle like a crumpled ball of paper to fit inside the skull. Compared to other animals, humans have more neurons in the cortex and a greater number of fibres connecting them to different brain regions. This interconnected network – which neuroscientists call the connectome – may be the key to our intelligence because it allows the rapid distribution of information around the brain. By linking regions involved in tool-making, imitation, social cognition, memory, causal reasoning and the senses, the brain could flexibly respond to changes in the environment. In other words, it could learn.

Although learning exists in the simplest animals, human learning is thought to have driven our brains' evolution forward by invoking something called the Baldwin effect. Named after the psychologist James Mark Baldwin, who described it in 1896, the theory says that individual learning enhances the overall learning of our species.

If, for example, an individual discovers a clever new strategy to feed the tribe, that discovery will offer any individual nearby a direct advantage. As more and more individuals thrive using the new strategy, they will reproduce more often and thus increase the chances of future generations making similar or even greater conceptual leaps. In this way, learning becomes installed in our genome not through the inheritance of DNA, but as a consequence of interactions with the environment. As the philosopher Daniel Dennett notes,

> It is possible for a behavioural innovation X to prove so useful that anybody who doesn't X is at a disadvantage, and variation in the population for *acquiring* or *adopting* behaviour X will be selected for

> genetically, so that after some generations, the descendants prove to be 'natural born Xers' who need hardly any rehearsal; Xing has been moved, by the Baldwin Effect, into their genomes, as an instinct.[4]

To prove that the Baldwin Effect was more than just an attractive theory, Geoffrey Hinton and Steven Nolan, computer scientists at Carnegie Mellon University in Pittsburgh, ran a computer simulation to show it in action. Using a genetic algorithm that learns by trial and error, they showed that over generations the genes that enable learning progressively increase in the population.[5] They conclude that the Baldwin Effect is an important process that allows humans to change the environment in which they evolve.

Putting it another way, intelligence didn't just encourage the proliferation of our species. It also encouraged the proliferation of culture. As humans became increasingly reliant on the cultural changes that enabled them to cooperate effectively – changing social patterns, changing social technologies, changing social goals – they became increasingly able to replace genetic evolution with cultural evolution. The cultural skills that humans acquired are why we've advanced more in the last 200,000 years than in the last 7 million. They set us on a trajectory of rapid technological innovation – the fruits of which, if tech experts are to be believed, will either save or destroy humankind.

It's also possible that other influences, working in tandem with cultural evolution, played a role in spurring our cognitive abilities to great heights. Among the most interesting is the idea that early humans may have eaten psychedelic mushrooms. Dubbed the 'Stoned Ape' theory, developed by ethnobotanist Terence McKenna in 1992, this idea carries more intellectual weight than one might think. Around 50,000 years ago, natural climate change forced early humans such as *Homo erectus* to abandon life in the forest canopy and take their chances out in the open. Had they remained in the canopy, humans today would have evolved to be a fruit-eating, insectivorous primate, and our evolutionary destiny would have been a lifestyle similar to the orangutans in Borneo and the lemurs of Madagascar. But as the African continent dried up, our ancestors found themselves flung from a bountiful supply of fruit and insects to an unforgiving desert with scant food and deadly predators. To survive, they did what they had always done: foraged for nutritious plants and juicy insects wherever they could find them. These insects were often found in the dung of wild cattle, and growing alongside was a variety of mushroom which, little did they know, contained the psychedelic drug psilocybin – 'magic mushrooms'.

It's possible that early humans, starving and desperate in the arid grasslands, ate these mushrooms and gradually changed the chemistry of their brains. Psilocybin improves something called 'edge detection': the brain's ability to determine the shape and profile of surrounding objects. More amazingly, it enlarges brain regions linked to learning, attention and creativity, and even increases the number of connections between them. The effect would have made our ancestors better hunters and survivors, swifter to ambush an antelope and quicker to spot a tiger in the bush.

Powerful stuff. But is it true? We know that psychoactive drugs were used in ancient cultures, including opium extracted from poppies in Neolithic Italy, hallucinogenic cacti harvested in around 8600 BC in Peru and marijuana found at the Bronze Age ceremonial sites in the Kara Kurum desert of Turkmenistan.[6] We also know that the remains of a prehistoric woman nicknamed the Red Lady (found in a grave in the El Mirón cave in Cantabria, Spain, and thought to belong to a thirty-five-year-old woman buried some 20,000 years ago) contained spores of several mushroom species lodged in her teeth.[7] This is tenuous evidence, to be sure, but the fact that psychedelics have such a potent effect on the brain means we should certainly be taking this idea seriously.

Wise Apes

No history of human intelligence would be complete without asking how we stack up against our closest relatives: chimpanzees and bonobos. It's a question that's divided anthropologists for centuries. Some maintain that humans are infinitely more intelligent than our ape cousins; others that the human brain is just a scaled-up primate brain – so cleverer, but not by much.

Let's start with the ape-flattering claim. The research on chimpanzees and bonobos suggests they are among the most intelligent beings on earth. They use simple tools and hunt in groups. They are social creatures, aware of status and capable of deception. They have been taught to use sign language and can do basic arithmetic with numbers and symbols. They even have culture, passing down customs from one generation to another. In 2016, scientists working in the Republic of Guinea observed chimpanzees engaging in ritualistic behaviour, piling rocks in the hollow of a seemingly venerated tree.[8]

Now for the unflattering news. Over the past two decades scientists have deployed numerous tests to measure the gap between human and chimpanzee cognition. What they find is that while both groups score equally well on tests of physical intelligence (tracking hidden objects, locating noise sources, using a stick as a reaching tool), they score differently on tests of social intelligence (solving problems through imitation, comprehending non-verbal cues, gauging someone's mental state by their behaviour). Time and time again, humans surpass chimpanzees on these tests by the age of four. The natural conclusion is that chimpanzees reach the levels of intelligence of a four-year-old child, but then hit their intellectual ceiling. Worse, if these tests are reliable they place the great apes' intelligence in the mid-Miocene epoch, 16 to 12 million years ago – ancient by anyone's standards.

But if our record of intelligence tests has taught us anything, it is to be deeply sceptical of them. According to the cognitive scientists David Leavens, Kim Bard and William Hopkins, there are problems with the way ape cognition has been measured over the years. First, in nearly every study the apes involved were institutionalised, born in captivity and isolated from their natural habitat. Captive chimpanzees display a range of abnormal behaviour that suggests mental illness. Second, the studies almost never age-matched the apes with the humans: infant humans were compared with adult chimpanzees. This makes it impossible to know if the test results are due to differences in intelligence or life history. 'The only firm conclusion that can be made,' Leavens and his colleagues wryly suggest, is that 'apes not raised in western, post-industrial households do not act very much like human children who were raised in those specific ecological circumstances, a result that should surprise no one.'[9]

Understanding the difference between human and other ape cognition is all about perspective. Does a chimpanzee need to understand human social life? No. Does a chimpanzee need to understand science, mathematics and philosophy? Of course not. Their particular environment does not require that particular adaptation. Similarly we do not understand the richness and complexity of chimpanzee life. We can use our science and observation skills to build an impression of their mental world, but that is all. The depths of their minds will never be fully understood from our perspective alone, and it is a colossal feat of human arrogance to think otherwise. Wittgenstein famously said that if a lion could talk, we could not understand her. Even if she spoke perfect English we wouldn't understand what she said because we share no common frame of reference, no common mental scheme by which to understand her world. He was right. Who knows what the world is like to a lion? Who knows what we are like to a chimpanzee?

So to answer the 'who's smarter' question: chimpanzees are staggeringly intelligent creatures that may be smarter than us in ways we have yet to discover. We just don't know. At least not until we develop more sophisticated ways of measuring intelligence. The primatologist Frans de Waal sums up our attitude to date:

> Having escaped the Dark Ages in which animals were mere stimulus-response machines, we are free to contemplate their mental lives. It is a great leap forward . . . But now that animal cognition is an increasingly popular topic, we are still facing the mindset that animal cognition can be only a poor substitute of what we humans have. It can't be truly deep and amazing . . . What a bizarre animal we are that the only question we can ask in relation to our place in nature is 'Mirror, mirror on the wall, who is the smartest of them all?'[10]

Misunderstanding Intelligence

One of the first to attempt to study human intelligence (from the Latin *intelligere*, translated from the Greek *nous*, meaning good judgement) was Francis Galton. In the late nineteenth century, statistician and polymath Galton maintained that intelligence was an inherited trait, akin to eye colour and height, which could be measured scientifically. He believed it could be tested using the grip strength and reaction times of English noblemen, and so instead of trying to measure things like knowledge, reasoning and creativity, he got thousands of volunteers to punch targets, distinguish between different colours and squeeze various objects. As they did, he measured their head size and recorded their academic record, determined to find, in his view, the 'very best people'. He coined the word eugenics (meaning 'the good birth') and encouraged people of the 'genius-producing classes' to go forth and multiply. It was a pernicious fallacy that formed the blueprint of Nazi ideology.

In 1904 the British psychologist and statistician Charles Spearman took another misguided step. He noted that people who perform well in one subject tended to perform well in another, seemingly unrelated subject. Children who scored highly on tests of vocabulary, for instance, were also likely to score highly on tests of arithmetic and vice versa. Spearman believed that this pointed to a mysterious underlying factor that he called General Intelligence, which he labelled *g*. In 1916 the American psychologist and eugenicist Lewis Terman used *g* to develop the intelligence quotient, or IQ test.

The problem with IQ, however, is that while it can help predict certain life outcomes – such as health and income – it's completely unreliable on the individual level. Spearman had merely taken a statistical correlation obtained from a small sample of British schoolchildren and given it the grand title of General Intelligence. But as modern researchers have pointed out, a more apt description would have been 'general test-taking ability'. Indeed, one of the greatest lessons I have learned as a neuroscientist is that intelligence is extraordinarily complex and subtle; it utilises countless neural circuits throughout the brain and often manifests in surprising ways. No single measure such as IQ can capture the diversity of cognitive ability that we see in our species: it's too simplistic.

Despite attempts to highlight Spearman's missteps – notably by the eminent psychologist Howard Gardner in the 1980s who pointed out that *g* totally ignores specialist abilities – *g* and IQ are still used as measures of intelligence today. Yet over the past century prominent scientists have repeatedly come to differing opinions on what intelligence actually is:

> Judgment, otherwise called good sense, practical sense, initiative, the faculty of adapting one's self to circumstances. To judge well, to comprehend well, to reason well, these are the essential activities of intelligence.
>
> Alfred Binet and Théodore Simon (1916)[11]

> The ability to undertake activities that are characterised by (1) difficulty, (2) complexity, (3) abstractness, (4) economy, (5) adaptiveness to a goal, (6) social value, and (7) the emergence of originals, and to maintain such activities under conditions that demand a concentration of energy and a resistance to emotional forces.
>
> George Stoddard (1943)[12]

> The aggregate or global capacity of the individual to act purposefully, to think rationally, and to deal effectively with his environment.
>
> David Wechsler (1958)[13]

> [The] process of acquiring, storing in memory, retrieving, combining, comparing, and using in new contexts information and conceptual skills.
>
> Lloyd Humphreys (1979)[14]

> The ability to deal with cognitive complexity.
>
> Linda Gottfredson (1998)[15]

This isn't to say IQ is meaningless. Some of the world's greatest geniuses including Paul Allen, Philip Emeagwali, Judit Polgar and Fabiola Mann have very high IQs. If you have a high IQ you're certainly more *likely* to have a high intelligence, but consider the Nobel Prize-winning physicist Richard Feyman, whose IQ was only slightly above average. Or Scott Aaronson, distinguished American computer scientist whose IQ is bang on average. For those who worry about having a low IQ, Aaronson writes, '[I]f you want to know, let's say, whether you can succeed as a physicist, then surely the best way to find out is to start studying physics and see how well you do.'[16]

As with many other conundrums in biology, scientists have tried to understand intelligence by categorising it. There are now various prefixes for intelligence – *emotional*, *social*, *fluid*, *crystallised*, *practical*, *analytic*, *interactional*, *experiential* and *perceptual*. Emotional and social intelligence are the most well-known and are fairly self-explanatory, the former loosely describing the ability to understand and perceive emotions (both one's own and others'), the latter broadly describing how well you interpret and respond to social interactions. Fluid versus crystallised intelligence essentially describes logic versus knowledge, respectively. Practical intelligence is probably the closest thing to common sense, namely the ability to apply ideas and demonstrate good judgement. An analytically intelligent person 'generally does well at school and on standardized tests, but is not necessarily creative or high in common sense,' explains Robert Sternberg, the psychologist who proposed it.[17] Interactional intelligence describes our relationship with tools and technology, how good we are at home DIY or mending a fuse, say. Experiential intelligence (another Sternberg invention) is how well we adapt to new situations and expand on new ideas, a form of intelligence closely linked to creativity. And perceptual intelligence, brilliantly explored by the physician Brian S. Boxer Wachler in his 2017 book, is our brain's way of distinguishing between reality and illusion, our ability to see past self-deception and observe the world for what it really is.[18] These distinctions have helped, but as we will see, we still have a long way to go in order to understand intelligence.

The Power of Imagination

Intelligence, it would seem, is hard to define, but is even harder to pinpoint in the brain. For years scientists thought that human intelligence came from the frontal cortex (the outer layer covering the front of the brain) because humans appeared to have a larger frontal cortex than other apes. And bigger was assumed to mean better; when Albert Einstein's brain was examined post-mortem, pathologists noted a larger than usual frontal cortex. But it turned out that scientists had based their interspecies comparisons on unscaled measurements: using total brain volume instead of in relation to body size. By this measure, llamas and sea lions were smarter than primates. Thanks to the work of Robert Barton and Chris Venditti at the University of California, San Diego, who correctly scaled the size of the human frontal cortex in 2013, we know now that our frontal cortex is actually nothing special relative to other primates.[19] So where does the staggering intelligence of people like Einstein come from?

The answer lies in our evolutionary roots. Just as our brains have much to grapple with in today's world of advanced technology, political unrest, and environmental catastrophe, our ancestors had to navigate a complex and ever-changing world, fraught with disease, climate change, predators, and tribal warfare. Gradually, through the process of natural selection, the brain developed increasingly sophisticated circuitry. And as that circuitry grew, it spawned something that Einstein said was more important than knowledge: imagination.

Since antiquity, our ancestors have been imaginative: wearing stone necklaces 100,000 years ago, carving on ostrich shells 85,000 years ago, adorning caves with animal paintings 15,000 years ago. Imagining that things exist can be highly advantageous. Large groups of people often collaborate best by attributing meaning to abstract ideas – such as gods, nations and money – that only truly exist in people's collective imagination. By thinking abstractly, human minds have taken things that exist only in our imagination and given them a form in the real world.

The ability to form images in the mind without direct input from our senses is typically viewed negatively: people with overactive imaginations are seen as daydreamers living in an imaginary world, overindulgent idlers constantly in need of being snapped back into reality. However, new research has found that a strong imagination and intelligence are linked. For the past thirty years researchers have been mapping something called the default network, a brain system that participates in daydreaming, mind wandering, reflective thinking and imagining the future. It turns out that while you daydream, your thoughts are free to wander into various domains of cognition, such as memory, experience, knowledge and visual imagery. People who engage in these cognitive practices therefore have greater access to the states of mind necessary to solve complex problems.

In the brain, the default network is a web of interacting circuits spanning the frontal, parietal and temporal lobes. It contains a number of 'hubs' and 'subsections' that are important for processing thoughts associated with daydreaming: thoughts about oneself and others, thoughts about experiences and goals, thoughts about purpose and determining the motivation of others – the inner dialogue with one's self that we often disregard as superfluous to intelligence and creativity.

Importantly, the default network is only active when a person is not focused on a task, when the brain is cycling through thoughts not associated with the immediate environment. This is in contrast with a system called the executive control network, a brain network responsible for controlling attention and awareness of the external environment. Although neuroscientists don't entirely agree about what the default network is doing when the mind wanders, some believe it is consolidating our experiences in order to make sense of our individual autobiographies, the narrative that defines who we are. Curiously, when people watch the same film, brain scans show that their default networks are in near-perfect sync with each other; the film's narrative has literally replaced their own.

Scott Barry Kaufman, director of the Imagination Institute at the University of Pennsylvania, likes to call the default network the imagination network.

> In the school system no one sees your inner stream of consciousness, your imagination. They only see how quick or slow you are at dealing with external tasks. But imagination is really our greatest

skill. It supports intelligence by allowing the combination of different types of knowledge or ways of thinking to create new ideas.[20]

In March 2016 researchers saw a glimpse of imagination in action for the first time: a bundle of specialised neurons, called grid cells, lighting up in the entorhinal cortex after volunteers were asked to imagine moving through a mountainous landscape.[21] Tellingly, this brain area is known to act as a hub connecting different parts of the brain.

We still don't know how these networks evolved, but many neuroscientists think they have something to do with an ancient brain function called repetition suppression. This phenomenon occurs when your brain becomes familiar with something, and displays less of a response each time it sees it. When the first iPhone was released in 2007, for instance, people's creative brain networks would have undoubtedly shown a large response; our minds were assimilating a novel, life-changing object for the first time. But having seen so many generations of iPhone, our brain's response is now much smaller. This mechanism probably evolved as a way to conserve energy, directing our attention to novel, potentially valuable new sources of information. For early humans, repression suppression may have been vital to improve their tools. Perhaps the diminishing novelty of sharpened stone tools 2.5 million years ago helped lead to stone hand-axes 1.6 million years ago, which subsequently led to stone knapping 400,000 years ago and cutting blades 80,000 years ago. Every improvement was the brain's way of keeping its creative juices flowing.

This need to keep the brain active was starkly demonstrated by a team of psychologists at the University of Virginia and Harvard University, who recently found that people resist being left alone with nothing but their thoughts, even for as little as six minutes – irrespective of age, education or income. Given the option, people actually prefer to receive a mild electric shock to being made to sit and do nothing.[22] Why our brains make this choice is a bit of a mystery, but we know that the brain needs stimulation in order to function well. A gentle electric shock can trigger the release of neurotransmitters at the synapse, activating the neural circuitry associated with reward. From an evolutionary perspective, this isn't surprising: our brains evolved to be active, to engage with their surroundings and seek out new information wherever they can find it, even if the cost involves momentary discomfort.

Although imagination starts in childhood, some believe it is educated out of us at school, where we are taught literacy, numeracy and ultimately conformity. Indeed, in the UK, US and elsewhere the school system is still based on the nineteenth-century Prussian model, in which children wear uniforms and are told not to challenge authority. Such strict obedience, neuroscientists say, can quash creativity. It does this by inhibiting our brain's dopaminergic system: a network of neurons that synthesise the neurotransmitter dopamine, which drives our motivation to explore the world and be creative. Too much dopamine can lead to fantastical thoughts and hamper the critical thinking necessary for creativity. But too little dopamine leads to unmotivated, poorly focused children who are then more likely to conform, generating a vicious cycle of dwindling imagination.

Kaufman is now investigating how teachers can promote imagination in the classroom. Some believe the key is to give schoolchildren time to reflect on what they have learned. In a 2012 report entitled 'Rest Is Not Idleness', the neuroscientist Mary Helen Immordino-Yang and her colleagues argue that introspection, quiet reflection and mindfulness improve academic performance.[23] For example, secondary-school pupils who write in a diary about their anxiety about upcoming exams get better marks. African-American pupils who take time to imagine themselves as successful adults do better at school. Academic accomplishments are even bolstered in ten- to twelve-year-olds taught to take what psychologists call a meta-moment: pausing in an activity to reflect and think about themselves at their very best. This isn't to say that pupils should indulge in unhealthy distractions – smartphone overuse in particular reduces default network activity, actually impairing the daydreaming state of mind. Rather, Kaufman and Immordino-Yang's research suggests that to start improving education, we must from time to time let our minds wander.

When scientists look to see what else triggers the default network, they discover something extraordinary. In one study, the pseudonymous university student John was told a true story designed to generate compassion. The tale was about a young boy growing up in China during a depression. The boy, fatherless and impoverished, had to watch his mother work as a labourer to survive. One day she finds a coin by the road and uses it to buy the boy warm cakes, which he offers to share with her despite being starving, but she declines. Asked how the story made him feel, John replied:

> This is the one [true story] that's hit me the most, I suppose. And I'm not very good at verbalizing emotions. But . . . um . . . I can almost feel the physical sensations. It's like there's a balloon or something just under my sternum, inflating and moving up and out. Which, I don't know, is my sign of something really touching . . . [long pause] It makes me think about my parents, because they provide me with so much and I don't thank them enough, I don't think . . . I *know* I don't. So, I should do that.[24]

The pauses in John's answer are manifestations of default network activity. What's interesting is how closely his pauses align with thoughts not pertaining to the immediate story. There are abstract thoughts – 'a balloon . . . under my sternum' – and reflective thoughts – 'my parents . . . I don't thank them enough.' The astonishing conclusion is that compassion stimulates the default network as well. How this happens has a lot to do with the brain's anatomy: the brain regions underlying compassion – the medial orbitofrontal cortex and the ventral striatum – which produce feelings of warmth, concern and tenderness are hugely interconnected with the default network. Therefore anything that activates these brain regions has the knock-on effect of activating default network activity too. Put another way, kindness makes you cleverer.

If this sounds unlikely, consider for a moment the evolutionary benefits of compassion: altruism, cooperation, generosity, reciprocity, forgiveness, self-sacrifice, love. Given that the human brain evolved by building on pre-existing circuits, it's no surprise that intelligence also springs from compassion. It's why we have emotional and social intelligence as well as the more familiar kinds. It's long been known that compassion requires imagination. 'Climb into his skin and walk around in it,' Atticus tells Scout in *To Kill a Mockingbird*. But the converse is also true: imagination requires compassion.

Why did evolution select such a surprising mechanism for intelligence? Why not just opt for a data-processing intelligence like that used in artificial intelligence (AI) software? The answer is simple: because imagination is far more powerful. Unlike AI, the human brain goes beyond analysing the environment and breaking down tasks into data. Instead, we constantly adapt to an environment that we have also shaped, gaining new knowledge while simultaneously learning from the past and imagining the future. We go beyond 'what is' to 'what could be'. We ceaselessly create new ideas; even the cells in our body ceaselessly create themselves.

In fact, the more we learn about human intelligence, the more we are realising it is nothing like AI. Compared to human brains, AI is clumsy and narrow-minded. An AI can quickly identify an object like a banana, for instance, but add a competing signal and the AI will suddenly think it's a toaster. If you came across a mountain that looks like an elephant, the AI would think the mountain *is* an elephant. AI also needs to learn a task thousands of times before achieving proficiency. And so, while AI outperforms humans in some ways – it recently outperformed doctors in diagnosing breast cancer, for instance[25] – it still doesn't re-create human intelligence. For all the thrill and prestige of AI, no machine is a match for the breathtaking ingenuity of human intelligence at its best. Seven million years of brain evolution have seen to that.

A Cleverer World

Since the beginning of the twentieth century humans have been getting cleverer. However we measure it, intelligence scores are going up around the world. If we could send a typical teenager back in time to 1900, he or she would be among the cleverest 2 per cent of people on the planet. Equally, if we brought a typical person from 1900 into the present, he or she would have significantly below-average intelligence. We call this phenomenon the Flynn Effect, after the New Zealand political scientist James Flynn who discovered it in the 1980s. And remarkably, it is almost certainly a product of the environment. Explanations include better health, better nutrition, better education and improved standards of living. But most significantly, the Flynn Effect proves that intelligence is changeable. As Flynn notes:

> The industrial revolution demands a better educated work force, not just to fill new elite positions but to upgrade the average working person, progressing from literacy to grade school to high school to university. Women enter the work force. Better standards of living nourish better brains. Family size drops so that adults dominate the home's vocabulary and modern parenting develops (encouraging the child's potential for education). People's professions exercise their minds rather than asking for physically demanding repetitive work. Leisure at least allows cognitively demanding activity rather than mere recuperation from work. The world's new visual environment develops so that abstract images dominate our minds and we can 'picture' the world and its possibilities rather than merely describe it.[26]

There are many reasons why intelligence is primarily a consequence of the environment, why it is not fixed at birth and why we are more in control of our intelligence than we think. The first reason is the effect of experience on the brain. Experience affects the formation and elimination of neuronal synapses: strengthening the brain in some areas, weakening it in others. We know this because lab animals housed in an enriched environment filled with toys, tunnels and opportunities for social interaction, consistently score higher on tests of cognitive performance than animals housed in barren enclosures. Experience in the form of practice also leaves its mark on the brain: London taxi drivers have a larger hippocampus than other people to help them navigate the capital's streets; violinists have larger brain regions controlling the fingers of the left hand. This enhancement and rewiring of the brain is most active during adolescence and early adulthood, but continues throughout adult life.

But to what degree does the environment alter the brain? In 2015 one group of researchers offered an answer, demonstrating that complex environments actually push brain evolution forward. Larissa Albantakis, a computational neuroscientist at the University of Wisconsin-Madison, has created a set of artificially intelligent computer characters – animated creatures she calls 'animats' – that possess a simple neural network.[27] She lets her animats play a video game in which they must catch falling blocks. The best catchers are then selected for a more advanced game in a more complex virtual world. After 60,000 generations, Albantakis has found, the animats not only become better video-game players, but also generate more complex wiring in their neural networks. They created a kind of digital neuroplasticity.

If the environment can change the mind of a computer in this way, imagine what it can do for the human brain, a computer that is still thirty times more powerful than our best supercomputers. Perhaps the most striking example of its impact on the brain comes from an unsettling set of experiments performed by the neuroscientists David Hubel and Torsten Wiesel in the 1950s. Working in a tiny basement laboratory at John Hopkins University, they sutured closed one eye of a kitten and then let the animal mature to adulthood. When the sutures were removed, they found that the cat was blind in that eye because its brain's visual cortex had not received enough visual experience from the outside world. Signals from the environment, not the brain, thus determine how the visual cortex is wired. Clearly, neuroplasticity makes the human brain acutely sensitive to its environment. This is good news: it means we possess far more control over things like emotion, intelligence and social behaviour than was previously thought.

A person's intelligence can also change across their lifespan. When children from deprived backgrounds are adopted and placed in financially stable families, they get cleverer. As with lab animals, being raised in an enriched environment – one that emphasises love, learning, achievement and personal growth – has a huge impact on cognitive development. Adults too can increase their intelligence by changing everything from diet to education to work environment. And while the jury is still out on whether brain-training games affect intelligence, we do know that human cognition is responsive to training.

Perhaps the most striking study showing that intelligence can change over time was conducted by Cathy Price and her colleagues at University College London. They found that the intellectual ability of teenagers in particular is far more malleable than once thought.[28] They picked nineteen boys and fourteen girls, aged between twelve and sixteen, and subjected them to brain scans and a range of verbal and non-verbal tests. They then repeated the tests four years later to see if their scores had shifted in any meaningful way. Remarkably, 39 per cent of the teenagers had an improved verbal score and 21 per cent showed improvements in spatial reasoning. When Price then looked at their brain scans, she saw that both results corresponded with growth in the left motor cortex (a region important for speech) and the anterior cerebellum (important for spatial cognition). Price notes:

> We have a tendency to assess children and determine the course of their education relatively early in life. But here we have shown that their intelligence is likely to be still developing. We have to be careful not to write off poorer performers at an early age . . . [they] may improve significantly given a few more years.[29]

Our beliefs also affect our intelligence. In 1968 the psychologist Robert Rosenthal and headmistress Lenore Jacobson showed that when teachers were told that some of their pupils were 'late bloomers', rather than underachievers, those pupils were transformed by their teachers' positive expectations and went on to perform significantly better than their peers. Simply believing in themselves was enough to change their lives forever. Rosenthal and Jacobson called this self-fulfilling prophecy the Pygmalion Effect, after the mythical Greek sculptor whose love for a female statue brought her to life.[30] The Pygmalion Effect has been reported in sports coaching, business management and medical training. People who believe intelligence can be improved can even perform better in maths tests than those who believe intelligence is fixed – the so-called entity theory. By adopting the idea that effort and determination is all it takes, they have the perfect attitude to succeed.

Sadly it is also true that when teachers have low expectations of their students, poor results are precisely what they get. Psychologists call it the Golem Effect and it is devastating. Children end up believing intelligence is fixed and perform poorly at school for no good reason. They become highly anxious about how clever they are and so aim low academically. They deliberately sabotage their chances of success (called self-handicapping) by not studying for tests, opting to play video games or watch television instead. In this way, when they fail a test their self-esteem is protected. The failure was their choice, not their fault.

Writing this resonates deeply with me because I experienced this very phenomenon. Growing up, I was taught that clever people were just born that way. You were either bright or a bit dim. There was no room for intellectual mobility; you were put in your lane and given a teacher set to the right speed setting. When I looked at the kids in the top class and those in the bottom class, I didn't see the outcome of nurture. I saw a predetermined hierarchy and it petrified me. I became so anxious I would retreat into myself, playing video games and messing about instead of studying for my exams. I went to a university ranked average for my undergraduate degree. But then something miraculous happened. One of my university tutors taught me that I could do so much better: that I could achieve whatever I wanted with perseverance and – most importantly – self-belief.

I listened. I stopped playing video games and studied hard. I got a first-class degree and, wanting to continue my studies of the brain, was accepted on a master's programme at University College London, one of the world's leading neuroscience institutes. From there it was a snowball effect. Surrounded by Oxbridge graduates, I suddenly believed in myself like never before. I worked even harder and outperformed most of them in the exams, earning a distinction and then a scholarship for a PhD – my ticket to neuroscience heaven and an interesting life that I could be proud of. None of which would have been possible without my tutor's kind and gentle Pygmalion influence.

And then there's the effect of motivation. It's easy to accept that motivation plays a role in intellectual achievement. Yet few people realise how far some cultures have taken it. Students from East Asian countries such as China, Japan and South Korea have a much stronger work ethic than white Western students. The academic motivation of Asian students arose from the teachings of the ancient Chinese philosopher Confucius (551–479 BC). Unlike Plato and other Western philosophers, who believed that intelligence is something one is born with, Confucius knew that intelligence is something one can earn. While Confucius acknowledged the existence of natural-born geniuses, he knew that such individuals are extremely rare. His focus on hard work has inspired generations of Asian students consistently to outperform their white counterparts in educational achievement.

Successes such as these remind us that for human intelligence to be advanced it must also be understood in a cultural context. The inescapable fact is that intelligence doesn't mean the same thing in every culture. In addition to motivation, Asians emphasise modesty, open-mindedness, curiosity and the sheer enjoyment of thinking. It's an almost child-like conception, bringing to mind Einstein's famous words in a letter to one of his colleagues: '[We] never cease to stand like curious children before the great mystery into which we are born.'

Other cultures, such as those in Latin America, emphasise the creative attributes of intelligence or, as in Ugandan culture, associate intelligence with slow and deliberate thinking. In Indian culture intelligence is linked to obedience, good conduct and following social norms, whereas in the West we typically associate intelligence with analytic thinking and placing objects into categories. This explains why many Westerners have had – and still do have – a one-dimensional view of intelligence. And I believe that, thanks to Spearman's *g* and our particular history of studying intelligence, our culture is segregated according to intellectual attainment. Our friends and colleagues are deemed more or less as intelligent as we are, and so our personal experience triggers a myopic attitude towards intellectual diversity.

The emerging picture of intelligence is one many of us already know. Intelligence is a muscle. All of us are born with the potential to use it, and it changes and gets stronger when we do. Very few of us – with the exception of rare geniuses like Srinivasa Ramanujan – are born either clever, of average intelligence or stupid. Like everything in the brain, intelligence evolved under ancestral conditions purely as an adaptation to change.

The effects of a good environment are not permanent, however. As soon as the nurturing influence is taken away from children, their achievements steadily diminish in what psychologists call the fadeout effect. In 2016, the University of California Santa Barbara psychologist John Protzko set out to discover how environmental interventions raise intelligence and exactly how long they last.[31] Protzko had the perfect test subjects: a group of 985 children who had been born with a low birth weight, and thus were exposed to an intense programme of cognitive training to make up for it. The training lasted until the children were three, at which point they were given a range of intelligence tests. Then, at ages five and eight, they were tested again.

Protzko found that the intervention had raised the children's intelligence at age three, but by age five the increases had completely faded. A mere two years of not exercising the mind to the same extent was all it took. Importantly, Protzko's study provided further evidence that a person's intelligence at one age has no bearing on their intelligence at another age. There is absolutely no causal connection, because intelligence doesn't work that way. It can be gained and lost as quickly as getting in or out of shape.

A crucial lesson can be learned from Venezuela in the 1980s, when a lawyer named Luis Alberto Machado decided to launch a bold three-year state programme known as Project Intelligence. An idealist and stalwart supporter of universal education, Machado believed – or rather knew – that intelligence is largely determined by experience and the environment. The project's goal was to teach thinking skills in seventh-grade classrooms (equivalent to Year 8 in England) in Venezuela. Teaching material included a range of tools designed to encourage problem solving, imaginative thinking, decision-making and self-confidence.

The results were astounding. Compared with children at six other state schools, pupils in the school carrying out the experiment scored nearly twice as high in a range of cognitive tests. To the tune of the mantra that every child has a right to an education, Machado argued that every child also has the right to develop his or her intelligence to the fullest. He published his views in a 1980 bestselling book, *The Right to Be Intelligent.*

The fact is, we all have the same neural hardware. What separates us from each other is the environment that we happen to have been born into. To say that a surgeon is cleverer than a bus driver would therefore be a misconception: the surgeon is simply using a different cognitive skill set, one that anyone could employ given the opportunity. This isn't an idealistic fantasy. The rise in educational achievement in the West has occurred over just three generations, much too fast to be the result of genetic change.

What we can see over the course of our history is a gradual awakening of the mind, which in living memory has been propelled forward at an exponential rate. In our evolutionary journey from tool-maker to atom splitter, in the elevating force of nurture over nature, in the power of beliefs and societal change, and in the enriching diversity of thinking types and cultural perspectives, we can see a new understanding of intelligence blossoming.

6

Creating Language

One evening in late 2018 Daniel Green, a sixty-eight-year-old bar owner from London, sat in his office with his twin brother and business partner, George. The siblings had opened their bar more than fifteen years before; Daniel had left his office job at an insurance company and George, a self-diagnosed loose cannon recently released from prison, was grateful for the new opportunity. 'It was our chance to start over,' George told me. 'We both had pasts that we'd rather forget, and our relationship had become distant, so I think opening the bar together was our way of putting things right. It was our happy ending.'

The pair were in the office for their weekly business meeting, a time to discuss everything from their stock, to operating costs, to designing new cocktails. As the more experienced of the two, Daniel usually did most of the talking. But this time was different. Daniel was quiet, confused and unresponsive. Then, apropos of nothing, he started babbling incomprehensible sentences – meaningless phrases about the 'time of the ocean' and 'the sound of appointments' – looking at George as if what he said was completely normal. Before George could work out what was wrong, Daniel collapsed. He was having a massive stroke.

At the hospital, the doctors explained to George that Daniel's stroke had occurred in the left hemisphere of his temporal lobe, damaging the language centres of his brain. He now has what's known as Wernicke's aphasia: a condition in which the production of speech is unaffected, yet the meaning of words is almost completely lost. Asked how his day is, Daniel will now babble, 'we stayed down the jango around the them there, up to spoodle with the family over the remrem towards the people, they're taking all the moment for us today, with Jack.' For a man who came top in his school spelling competition with the word 'cymotrichous' (having wavy hair), it's a devastating loss.

In July 2021 I video-called Daniel and George at their flat in South London – Daniel has curly grey hair, green eyes and a polite, self-effacing demeanour; George resembles his brother, but is taller and looks less frail. They'd agreed to the interview to raise awareness of aphasia – an isolating disorder that many are unfamiliar with – and also to learn if I knew of any advances in research in to how the brain processes language. Naturally George led the discussion, acting now as a carer as well as a devoted brother. He said he often knows what Daniel is trying to say, and thinks of his role more as a guide than anything else. His brother's mind was intact, but its voice lost at sea. And George was the navigator.

I asked Daniel how he was, and what he'd been up to. 'You know that polkamoot dazzled with our friend the other day and we want to take them up the hill to see fans created by lightning,' Daniel said. 'It's hard to raiserbad the flower pots since we went up there.'

I smiled and nodded and looked at George, unsure whether at least to try to respond. 'That's Daniel's way of saying we went to a barbecue yesterday,' said George.

> 'Polkamoot' is new. It might be something to do with the children there. One was wearing a Pokemon T-shirt. The 'fans created by lightning' is probably something to do with the barbecue itself. As for the 'flower pots', your guess is as good as mine.

Daniel chimed in: 'He gets the river for the time we all need. Them, the tea over then.' 'Oh that's an easy one,' George said. 'He's just asking me to go make some tea. You can see how it's manageable. He just needs someone with him a lot more now.'

While George stole out to make the tea, I asked Daniel questions about his life and career running the bar. I asked him how he felt about his condition and how it had changed his everyday life. His answers, though nonsensical to me, were fluent and thoughtful; he was clearly of sound mind in other aspects. At times, it looked as though he was trying to find the right words, even though most patients with Wernicke's aphasia don't know how their speech is coming across. According to George, Daniel usually thinks he's making sense but then becomes frustrated when it's clear that he isn't.

When George returned, the conversation turned to deeper musings, and how Daniel is now involved in research projects to help scientists understand how the brain creates language. By using fMRI to map the language areas of his brain, scientists can examine which areas have remained intact, and whether or not they contribute to tasks such as verb generation, object naming and sentence comprehension. It's still unclear which of these came first in evolutionary history but, by examining Daniel's temporal lobe, scientists hope to discover the brain circuitry responsible. If they succeed, they can then look for signs of that circuitry in other primates, such as bonobos and chimpanzees, and thus help expose the evolutionary origins of language. In the epic story of human brain evolution, language was a pivotal moment.

So when did humans start talking to each other? Estimates range from 50,000 to more than 1 million years ago, for words leave no trace in the fossil record. Why did we start talking? Some think it began with our ancestors imitating various natural sounds, others that it arose from spontaneous cries of pain or surprise, or that it evolved from grunts, groans and snorts prompted by arduous physical labour. The theories are as imaginative as they are disputed.

Howsoever it arose, language became the cornerstone of human cognition. Over 7,000 languages are spoken in the world today. While other animals communicate using sounds, smells and body language, humans have gone further, creating tens of thousands of arbitrary symbols to build communities, cultures and societies. As we shall see, the stories we tell ourselves are merely adaptations for social interaction. Having evolved for emotion, sociability, memory and intelligence, the human brain could next evolve a system to connect us all.

Language probably began over 1 million years ago with *Homo erectus*, a supremely clever early human that made tools, lived in groups and built boats to colonise islands such as Flores in Indonesia, and Crete. '*Erectus* needed language when they were sailing to the Island of Flores,' says the American linguist Daniel Everett.[1] 'They needed to be able to paddle. And if they paddled they needed to be able to say "paddle there" or "don't paddle". You need communication with symbols, not just grunts.' Everett accepts that such language would have been no way near as sophisticated as that used by modern humans; nonetheless, he believes that *Homo erectus* invented symbols capable of initiating the baby steps required for grammar and syntax. In other words, there was no single moment when language crystallised. It emerged gradually when small bands of people decided to adopt culturally invented symbols.

Scientists call this the sign progression theory of language. Symbols (invented markings that represent something, such as an animal) arose from icons (pictures that physically resemble the thing, such as a cave painting of an animal) which themselves arose from indexes (something connected to the thing, such as a footprint of an animal). Over time, ancient symbols would have been combined with others to produce increasingly complex grammar and sentence structures. The language we speak today is therefore the legacy of all the words ever uttered by our species. From a long line of language-acquiring ancestors and a rich pattern of cultural inheritance, language has flowed through time like water through a canyon.

The evolutionary value of language is hard to overstate. 'As soon as a system of symbolic communication came into being,' the biologist Jacques Monod once wrote, 'the individuals, or rather the groups best able to use it, acquired an advantage over others incomparably greater than any that a similar superiority of intelligence would have conferred on a species without language.'[2] Or as the psychologist Cecilia Heyes put it, 'Now capable of abstract thought and subtle communication, we became radically different from all other animals – more like gods than beasts.'[3]

To be sure, animals perform dazzling feats of communication. Octopuses, lacking vocal cords and unable to emit any sound, send signals with their skin colour and texture. Squid display bands of colour to produce an alphabet of patterns, said to have at least fourteen 'letters'. Moths use pheromones, birds click their wings, dolphins whistle, honeybees dance, elephants caress with their trunks. Some animals even communicate with other species – in exchange for a share of the spoils, the honeyguide bird leads honey badgers to bees' nests by making a distinctive trilling sound.

We may never know definitively when the first human language appeared, but it's likely that all languages share a common origin. That origin could be the result of a bottleneck effect in which a single language then became ancestral to all the world's languages. 'This language would have been spoken by a small east African population who seemingly invented fully modern language and then spread around the world, replacing everyone else,' says the linguist Merritt Ruhlen, who has searched for clues to the first human language by creating a phylogenetic tree showing the historical relationships between all the language families in the world.[4] One family is the Indo-Aryan family (which includes Bengali, Hindi and Punjabi), which is a branch of the Indo-Iranian family (which includes Persian, Pashto and Kurdish), which is itself a branch of the even larger Indo-European family, said to be between 6,000 and 8,000 years old, and which includes languages as diverse as Danish, Greek, Hindi and Welsh.

Of course, language is more than a collection of symbols. It's also an inventory of sounds (phonetics) coupled with an inventory of sound structures (phonology). Early humans exploited this fact in a remarkable way. Researchers have found that ancient cave art is often located in acoustic hot spots: spaces where sound generates echoes. These cave chambers are harder to reach than chambers that have no echo, so it's thought they were deliberately chosen as places for secretive rituals and dramatic storytelling. Strikingly, when acoustic archaeologists study the sonic properties of these caves, they find that paintings of cloven animals such as bulls, bison and deer, appear in chambers that generate echoes and reverberations that actually sound like hoof beats. Quiet chambers, on the other hand, are painted with dots and handprints. The finding is now so consistent that some acoustic archaeologists can locate the animal paintings in pitch black, the echo of their voices being their only guide.

When our ancestors first used language they had no idea how it would shape their brains. But just as exercise changes our bodies and diet changes our gut, language does indeed change the physical structure of our brains. Neurons form durable connections in response to sights and sounds – including words – and this neuronal wiring forms the basis of the brain's grey matter, where all information about the world is processed. As children, we pick up words and phrases unconsciously, because the sounds of our native language strengthen certain neuronal connections. For monolingual people, this means that some neuronal connections are weakened due to certain sounds being absent in their language. Which is all the more reason to teach children multiple languages: the additional input actually enhances the physical structure of their brains. Bilinguals use their anterior cingulate cortex (ACC) – a brain region important for empathy, emotion and decision-making – in a more efficient fashion than monolinguals. They possess a larger auditory cortex, making it easier for them to detect changes in pitch, locate sounds in space and understand new languages.

They even encode speech sounds in their brain in a more robust manner, giving rise to a richer phonetic environment.

Changes in one brain area nearly always trigger changes in another. Remarkably, the increased brain activity of bilingualism can delay the onset of dementia by around four years, as well as help the brain cope with dementia after developing the disease. The additional neuronal pathways are also thought to enhance cognitive skills such as memory, attention, multitasking and creativity. Geographical constraints meant that our ancestors were unlikely to be learning several languages at once, but the important thing to remember is that a second language merely amplifies the cognitive benefits that occur in monolingual people. Language does for the brain what the latest software update does for your computer – it makes it faster, cleverer, more agile and better at completing tasks. For early humans, language was the software update that made better thinking go viral.

Why Language Evolved

So why did language evolve? Was it, as acoustic caves suggest, for storytelling? Perhaps. The desire to tell and hear stories knits a community together. They foster a sense of shared values, of togetherness and cultural cohesion. Fictional narratives are especially important to society: as humans become increasingly dependent on one another, as we evolve to live in ever-bigger societies, we need such narratives to learn how to cooperate. When anthropologists visited eighteen groups of hunter-gatherers in the Philippines in 2017, they found that 80 per cent of their tales concerned morality and social justice.[5] The most skilled raconteurs also proved to be the most desirable social partners, even more so than skilled foragers.

Or perhaps language evolved from shared intentions – that is, the power of human minds to collaborate to achieve a collective goal. In this context, language is merely an expression of the mind's ability to unite with others, a veil concealing more basic objectives that evolved in tandem with cultural progress. This view is gaining significant ground in the academic community. Since languages change more rapidly than genes, this brain change was shaped mainly by culture, not biology. Consider the Romance languages (French, Italian, Spanish, Portuguese, Romanian): they diverged from Latin in less than 2,000 years. No evolutionary change works that fast.

These observations teach us arguably the most important lesson about why language evolved: it's a social innovation. A collective act to facilitate human cooperation. Just imagine a world in which we had not evolved the capacity for language. Instructing the young and passing on knowledge of the natural world would have been very basic, not to mention building villages, planting crops and developing new technologies. As humans evolved, language became almost inevitable: with increasing contact between bands of early humans came an increased need to establish cross-group communication networks. Material culture was especially important. Trade creates opportunity and 'with trade comes negotiation and further selection for effective communication,' writes the anthropologist John Odling-Smee.[6]

Cultural evolution propelled language forward with one of the greatest inventions of all time: the printing press. In 1440, when Johannes Gutenberg first used the imposing wooden structure for books, newsheets and other printed knowledge, so too were new words and new ideas distributed throughout the world, radically expanding each language's vocabulary and changing how thoughts were shared. The spread of science and technology in particular – much of which was based on Greek texts that were preserved by Arabic and Persian translators – led to the intellectual movement widely known as the Enlightenment. Before the fifteenth century, the world's literacy rate was below 10 per cent. Today, it's 83 per cent (91 per cent for young adults aged fifteen to twenty-four).

Surprisingly, given how quickly we've adapted to the task, reading and writing are not things brains are born to do. Certainly no hard-wired brain structure evolved for such tasks, so how does the brain do it? In 2012, neuroscientists at Stanford University showed that reading ability in young children is linked to the growth of white matter tracts in the brain.[7] White matter – named after the white fatty substance (myelin) that surrounds nerve fibres – allows the rapid transmission of information throughout the brain. Importantly, studies also found that practising reading can boost white matter integrity, improving children's reading ability and their capacity to learn more generally.

Reading changes the brain by changing its connectivity. When we read, connectivity in the left temporal cortex and central sulcus – areas important for language and the sensation of movement, respectively – is increased. Which is why reading a novel can transport you into another world. If a character you are reading about is running, for example, neurons in these areas boost their connectivity and start mimicking the physical sensation of running. Neuroscientists call this phenomenon embodied cognition, and it's a way for our brains to incorporate bodily experiences into how we think.

In evolutionary terms, this is a major advance because it circumvents the fact that different areas of the brain have adapted for different tasks. By ramping up the brain's connectivity, reading forces the brain regions important for attention (the temporal lobe), learning and memory (the hippocampus), and object recognition (the fusiform gyrus) to work together to produce something novel. The fusiform gyrus also helps us distinguish between words and letters, for instance. Which is pretty impressive, given that *Homo sapiens* have only been reading for 5,000 of our 400,000 years of existence. It's an exceedingly new activity, and it will probably take us a long time fully to discover how it is changing our minds.

Human minds have certainly been eager to exploit language. Since modern humans first evolved, as many as 31,000 languages have come into existence. Much of that linguistic diversity has sadly perished. Of today's dwindling 7,000 languages the ten most spoken – English (1.13 billion speakers), Chinese (1.11 billion speakers), Hindustani (544 million speakers), Spanish (527 million speakers), Arabic (422 million speakers), Malay (281 million speakers), Russian (267 million speakers), Bengali (261 million speakers), Portuguese (229 million speakers) and French (229 million speakers) – lay claim to half the world's population. Many of the rest are concentrated in only a handful of locations. One in eight of all the world's languages are spoken by tribes in Papua New Guinea, thought to be due to a unique landscape that allows small groups to flourish independently. In the village of Ayapa, Mexico, the pre-Colombian language Ayapaneco almost died out because the last two surviving speakers refused to talk to each other.

The main factor affecting linguistic diversity, though, is not culture but climate. The anthropologist Robert Lee Munroe observed that people in warmer countries rely more on vowels because they're easier to hear outside than consonants. Speakers in colder climates, on the other hand, use more consonants because most communication occurs indoors. It's called acoustic adaptation, and is seen across the animal kingdom. Birds such as sparrows, blackbirds and the great tit sing at higher pitches in cities in order to penetrate the noise produced by traffic and human activity. Woodland birds sing at lower frequencies than birds living in the open in order to cut through branches, which deflect high-frequency sounds. In 2015, the linguist Ian Maddieson and his colleagues took 633 languages and explored the ecology and climate of where each language developed.[8] (Languages such as English, Mandarin Chinese and Spanish were excluded from the study since they are no longer restricted by geographical location.) The pattern that emerged gave resounding support to Munroe's observations: languages in hotter, humid climates were vowel-rich and mellifluous; languages in colder, drier climates were consonant-rich and guttural.

How We Learn Language

Social experience is crucial for language development, hauntingly demonstrated by the existence of feral children: children who grow up with no human contact and thus no experience of human language. Such children – usually confined by their own parents – can only speak a few words and find it tremendously hard to learn a language after being isolated for so many years.

There have been over a hundred reported cases of such children. There was Oxana Malaya, a Ukrainian girl whose parents left her in a dog kennel behind her house. When the authorities found her five years later, she couldn't speak, ran on all fours and barked. There was Victor of Averyon, a French boy who lived alone in the woods in the late 1700s. At twelve years old, he was caught by local hunters and found to speak no recognisable language. And there was Prava, a seven-year-old Russian boy who communicates only by chirping after his mother raised him as a pet bird in her aviary. Dubbed 'bird-boy', Prava doesn't understand any human language and flaps his arms when he knows he is not understood. But perhaps the most famous case of all is that of a Californian girl nicknamed Genie.

Genie was born in 1957 to dust bowl migrants from Oklahoma. Her father, a machinist who worked on an aircraft assembly line during the Second World War, didn't want children and hated Genie from the start. A deranged and tyrannical man, he labelled her retarded when she was twenty months old and decided to keep her as isolated as possible. For the next thirteen years he kept her either strapped in a straitjacket that was tied to a toilet, or bound in a wire-mesh covered crib with her arms and legs restrained. If she cried or made a sound, he would growl at her like a dog and then beat her with a baseball bat.

Genie's nightmare existence may never have been apprehended were it not for a chance occurrence one day in October 1970. Her mother – a loathsome figure who was mostly indifferent to her daughter's plight – had cataracts and decided to take Genie with her to apply for disability benefits for the blind. Because of her poor eyesight, though, she accidently stumbled into the wrong office: the office for social services. Upon seeing a hunched, withered and shaking girl beside her – a girl they guessed was six but were shocked to learn was in fact thirteen – the social workers immediately called the police. When Genie's father was charged with child abuse, he shot himself. 'The world will never understand,' he said in his suicide note.

Genie was moved to the Children's Hospital Los Angeles. She weighed just fifty-nine pounds, had deformed hips and an undersized rib cage. She was unable to straighten her arms and legs, and moved with what the nurses called a 'bunny walk', holding her hands in front of her like claws. According to one doctor, she was 'the most profoundly damaged child I've ever seen.' Her rehabilitation team included paediatricians, psychologists, linguists and scientists from around the world. Among them was Susan Curtiss, a linguist determined to teach Genie English. Although Genie could speak a few words, she couldn't produce grammatical sentences. A transcript of one of her first attempts reads: 'Father hit arm. Big wood. Genie cry . . . Father hit big stick. Father angry. Father hit Genie big stick. Father take piece wood hit. Cry. Me cry.'

After a few months, Curtiss and her colleagues saw improvements in Genie's vocabulary. She could learn new words and would occasionally put two or three together. 'She was smart,' said Curtiss. 'She could hold a set of pictures so they told a story. She could create all sorts of complex structures from sticks. She had other signs of intelligence. The lights were on.'[9] To psychologists, Genie's case was the perfect chance to settle a debate. There are two models for language development: the nativist and the empirical. The nativist posits that language is biological and depends on circuitry installed in the brain at birth, a so-called 'language acquisition device' that allows children to learn language quickly and easily.

The empirical model describes language as a product of the environment, with a 'critical period' of learning that begins at birth and lasts until the onset of puberty.

Genie's inability to learn language proved the latter to be true. In the years following her release, separate research showed that a child's brain develops much of the machinery for language before the age of ten: regions of the auditory cortex connect to the frontal lobes to sprout pathways for speech production, speech recognition, phonological memory and lip reading. The ability to learn language fluently, to apprehend its grammatical complexity seamlessly, persists until the age of eighteen and then tails off in adulthood. This doesn't mean that learning a new language in adulthood is futile, but it's hard, and Genie's extreme isolation and abuse made it impossibly so.

This is because the brain becomes fine-tuned for language upon contact with others, an idea first proposed by the Soviet psychologist Lev Vygotsky in 1930. As far as public intellectuals go, Vygotsky was as influential as Sigmund Freud and Ivan Pavlov. The Communist Party censored much of his work and so he only became known to Western scholars in the 1970s. In publications spanning a ten-year period, Vygotsky made the following novel argument: language – indeed much of human development – evolves from the interaction between individuals and society. Society can change people as much as people can change society. Children learn language using what he called a 'zone of proximal development', a gap separating knowledge from ignorance that can easily be filled by a 'more knowledgeable other': a teacher, a parent, a friend. 'Through others,' he wrote, 'we become ourselves.'

Today we live in a world that takes Vygotsky's reasoning for granted. Qualified teachers, good school attendance and sound parental care are enshrined in law. Yet Vygotsky's claim that other minds literally sculpt our own when learning language is one of the most staggering ideas in neuroscience. It suggests that language evolved to link our brains together physically: to create, in effect, a hive mind. Like the densely packed society of neurons in our head, each communicating using a chemical web of neurotransmitters, language binds human societies together using a collection of sounds, symbols and meanings. Language is the interconnected web of experience writ large.

In fact, society changes language in a process very similar to evolution by natural selection. Words, like genes, survive and produce offspring based on how adapted they are to the environment. The word aspirin for instance was invented in 1899 by a German chemist and derives from the opening letters of Acetylirte Spirsäure (acetylated spiraeic acid); it survives because it's easy to say. Shakespeare, a prolific neologist, invented words such as critic, swagger, lonely and hint. Other new species include fragrance and pandemonium (John Milton), universe and approach (Geoffrey Chaucer), valediction and self-preservation (John Donne), anticipate and atonement (Sir Thomas More). English speakers are adding a thousand new words a year to their lexicon. Some of my favourites include blog, doublethink, quidditch, podcast and McJob. The word selfie has even spawned its own offspring: there's helfie (a selfie of one's hair), welfie (a selfie during a workout) and drelfie (a selfie when drunk).

Words, like species, can also go extinct. In 2009 the biologist Mark Pagel and his colleagues at Reading University used supercomputers to apply the theory of natural selection to the family of Indo-European languages. The results suggested that a number of commonplace words such as 'dirty', 'bad', 'turn', 'guts', 'throw' and 'stick' may become extinct or be permanently replaced by new words over the next 700 to 1,000 years.[10] Pagel's analysis showed that the most stable words were those essential to daily conversation, such as 'I', 'we', and the numbers one, two and three. Adjectives and adverbs were found to be particularly susceptible to dying out. In fact, half of the words we use today, Pagel points out, would be unrecognisable to our ancestors 2,500 years ago. As the noted philologist Max Müller observed in 1870, 'A struggle for life is constantly going on among the words and grammatical forms in each language.'

The Internet is almost certainly accelerating this process. From fashionable punctuation to emojis to repurposing other symbols (the # was recently named UK children's word of the year), the way we talk online is changing fast enough to make even the most progressive baulk. Trying to stop this is futile. In a recent study, Joshua Plotkin, a mathematician and biologist at the University of Pennsylvania, charted how grammar changes over long stretches of time.[11] By examining over 100,000 texts published over the past 200 years, Plotkin and his team found that languages evolve by random chance as well as natural selection. An example of natural selection would be when verbs are regularised: the word 'smelt' changing to 'smelled'. They're easier to learn, thus more adapted to their environment. Verbs changing by random chance are those that became irregular: 'dived' changing to 'dove', 'lighted' to 'lit', 'waked' to 'woke'. There's no reason they should have changed; they just have. In the natural world this kind of change is called genetic drift, the second major force of evolutionary change, when genes become more or less frequent by pure fluke. And so if, like me, you want to move with the times but also keep some of those cherished rules of grammar and syntax, prepare to be disappointed.

As children, we practise language through pretend play. From about eighteen months onwards, all children, in all cultures, engage in pretend play: having imaginary conversations with made-up characters, talking into a banana as if it were a telephone. Childhood psychologists encourage pretend play because it appears to have a positive effect on a child's ability to learn. The pretence is often full of rich collaborative dialogue and set in an environment, usually the home, that reduces the pressure of learning. Additionally, neuroimaging studies show that the same brain circuitry underlying childhood pretend play is essential for adult creativity. Neurons in the prefrontal cortex, dorsal and ventral striatum motivate children to engage in pretend play; and then, as adults, the same population of neurons become vital for creative thinking. The link between language, pretend play and creativity makes perfect sense: language is built on social rules, and when children pretend to be something else, they are experimenting with the rules, thinking creatively to dream up novel visions for the future.

The Speaking Brain

None of this is to say that biology isn't important for the development of language. In the relentless nature versus nurture debate, it is accepted that language evolved as a composite of both. Several brain regions are responsible for language. The most studied are Broca's area (left hemisphere, frontal lobe) which is important for speech production, and Wernicke's area (left hemisphere, temporal lobe) which is important for speech comprehension. Less understood are the angular gyrus and insular cortex, regions thought to unite the essential supplements of language – hearing, vision, attention, emotion and motor control – into a cohesive whole. Strikingly, none of these regions is dedicated exclusively to language. Some can even transfer their roles to other regions; Broca's area can be destroyed without affecting language, for instance. The borders within the human brain are dazzlingly blurred.

Some believe the secret of human language lies in a region within Broca's area: a bundle of nerve fibres called BA44, which is underdeveloped in chimpanzees, weak in new-borns, but fully formed and strong in adults with language ability. 'This fibre tract,' says Angela Friederici of the Max Planck Institute of Cognitive Neuroscience in Germany, 'could be seen as the missing link which has to evolve in order to make the full language capacity possible.'[12] We don't yet know if *Homo erectus* or another ancestor had BA44. Many researchers, therefore, hold to the notion that Friederici's thesis will have to remain a striking observation.

At the genetic level, scientists have long been intrigued by the *FOXP2* gene in language evolution. *FOXP2* made international headlines in 2002 when researchers found evidence that a special version of the gene spread to all *Homo sapiens* about 200,000 years ago. Songbirds use a version of *FOXP2* for vocal learning, as do chimpanzees, whales, dolphins and bats. FOXP2 controls the muscles used in speech, and mutations in the gene can lead to severe speech and language problems. One example is a condition known as apraxia, which affects the sequence of sounds, syllables and words by damaging the brain regions controlling the movements of the mouth, namely the parietal cortex and the corpus callosum. The parietal cortex, remember, is one of the four major lobes of the cortex, sitting at the back of the skull, just behind the frontal lobe. The corpus callosum (or 'tough body' in Latin) is the thick bundle of nerves separating the entire left from right hemispheres.

It's thought that human predecessors used a more ancient *FOXP2*, meaning they had less speech control, thereby giving *Homo sapiens* the evolutionary upper hand. This seductive narrative ran aground in 2009, however, when scientists discovered that Neanderthals also had the human form of *FOXP2*. That pushed our possible language gene back by at least 700,000 years. Did Neanderthals speak like *Homo sapiens*? It's certainly possible as Neanderthals did have an anatomically human hyoid or lingual bone – a horseshoe-shaped bone in the neck crucial for speaking. So they were probably capable of speech. Whether they actually spoke or not is a mystery.

Some scholars think that language evolved not from vocal calls, but manual gestures. Dubbed the gestural theory, the idea is that primitive sign language slowly became integrated into the circuitry of the brain, which then freed the hands for other functions. Eventually, the brain organised this sign language into a hierarchy of phonemes that were grouped together to form words, which formed the basis for sentences. Is there evidence for it? Absolutely. Primates use gestures to communicate with each other: in 2019 video footage of chimps living in Uganda's Budongo Forest recorded fifty-eight types of playful gestures. The most frequently used gestures were brief, staccato-like bursts, while longer sequences comprised short, syllable-like gestures. This may be why children use gestures to communicate before they have the ability to speak: our brains are hardwired to gesture by evolution.

In Nicaragua during the 1980s, a community of deaf children offered perhaps the strongest evidence for the gestural theory. The children, born to parents who could hear and having had little contact with each other until the government opened a school for the deaf in 1977, each had their own simple sign system at home and thus needed to develop a common sign language when brought together at the school. Which is precisely what they did. Over the course of thirty years, successive generations of children have created an entirely new sign language out of hand gestures, known today as Nicaraguan Sign Language (NSL). It was learnt unbidden, with no influence from any existing language, yet has rules found only in other languages. What surprised scientists the most, though, was how closely it followed evolutionary logic: the first signals were crude, producing only a few gesture traits that survived into the second generation, who in turn produced more sophisticated gesture traits that survived into the third generation, and so on.

The Nicaraguans aren't alone. In the Negev desert of southern Israel, some 150 deaf members of the al-Sayyid Bedouin tribe are in their seventh generation of evolving their own sign language, the al-Sayyid Bedouin Sign Language (ABSL). Stretching back over seventy years, ABSL has a longer history than NSL and evolved in a more natural setting than a school. But like NSL, ABSL is a totally unique language. It shares no characteristics with Bedouin Arabic, modern Arabic, Hebrew or Israeli sign language.

And it possesses a curious complexity in that it combines simpler words to achieve a different meaning: 'sweat' and 'sun' mean summer, for instance; 'pray' and 'house' mean mosque. Tribe members use ABSL to discuss everything from their dreams and desires to national insurance and local construction projects. Linguists are now keen to see how (and if) the language continues to evolve; unfortunately, the tribe has become isolated because it is stigmatised by other Bedouin tribespeople.

Two crucial motor systems evolved to enable speech: the mesencephalic periaqueductal gray (PAG) and the caudal medullary nucleus retroambiguus (NRA), both of which control neurons involved in vocalisation and our ability to modulate vocalisation into words and sentences.

The brain's PAG generates human speech in a mysterious way. Occupying a column of brainstem roughly half an inch long, the PAG can elicit vocalisations in animals when it is poked and prodded in a laboratory setting. Experiments on (I'm sad to say) cats found that four different sounds emerge when the PAG is electrically stimulated – mews, howls, cries and hisses – depending on where in the PAG the stimulation occurs. In humans, lesions in the PAG result in absolute mutism; we completely lose our ability to produce sound.

While the PAG's underlying neurophysiology remains unclear, many believe that it somehow weaves together breathing and laryngeal muscle patterns to generate a code essential for speech and song. Remarkably, that code was recently decoded into written text by researchers at the University of California, San Francisco.[13] Using a group of epilepsy patients about to undergo surgery for their condition, the researchers placed electrodes directly onto their brains and asked them to read out a list of twenty-four responses to nine sets of questions. This data was then used to build a computer model that matched patterns of brain activity to the patients' spoken words. Though the accuracy wasn't perfect (76 per cent) and the technology still fairly basic, the researchers now want to attempt to decode 'imagined speech', the inner voice that only we can hear.

Much less is known about the brain's NRA, only that it receives signals from the PAG and then sends its own signals to the diaphragm and other muscles controlling breathing. Unlike the PAG, the NRA does not generate speech patterns. Each of its neurons produces a specific motor action or tone, combinations of which can assist vocalisation, vomiting, coughing, sneezing, even mating posture and child delivery. The brain chooses the action based simply on what is needed in the circumstances. In this way, the NRA has been likened to a piano and the PAG the piano player. Together they exhibit a little-understood brain change that was critical for human speech.

In fact, it's now thought that monkeys and other apes cannot speak because they didn't evolve the PAG and NRA. Given how closely related we are, one would think that our primate cousins would at least be able to outspeak a parrot. For a long time it was thought that monkeys' and apes' vocal anatomy meant that they could not: if only they had more flexible vocal cords and larynx muscles, they would be able to mimic some human speech. Darwin wasn't so sure. He thought apes couldn't talk because of the architecture of their brains; and he was right. In 2016 a team of researchers led by Tecumseh Fitch at the University of Vienna and Asif Ghazanfar at Princeton University in New Jersey filmed long-tailed macaques with an X-ray scanner while they made various calls and sounds.[14] Then, using X-ray stills from the video, they made a computer model to show all the sounds the monkeys could make when air flows through their vocal anatomy.

They found that the monkeys do, in theory, have the anatomy to produce vowel sounds, including those in the English words 'bit', 'bet', 'bat', 'but' and 'bought', and the vowel-heavy phrase, 'Will you marry me.' The simulated monkey voice sounds a little artificial, but it's undeniably intelligible. 'A monkey's vocal tract would be perfectly adequate to produce hundreds, thousands of words,' Fitch said in an interview.[15] So why don't they? Because they lack the proper wiring in their brains; specifically, the wiring needed for the PAG and NRA. 'In short,' Fitch and Ghazanfar conclude, 'primates have a speech-ready vocal tract but lack a speech-ready brain.'

Not everyone agrees that monkeys have speech-ready vocal tracts, however. Philip Lieberman, a cognitive scientist at Brown University, Rhode Island, has argued that even if monkeys possessed the cognitive hardware to speak, their vocal range is simply too low to produce human-like speech. 'If monkeys had brains capable of learning and executing the motor commands involved in human speech, their "monkey speech" would not be as robust a means of vocal communication as that of fully modern human beings,' he writes.[16] Part of the problem is that monkeys cannot produce the 'quantal' vowels (i, u and a, and vowels in the words 'see', 'do' and 'ma'), which are present in nearly all human languages and are thought to be critical to the robustness of human speech. Fitch and Ghazanfar retort that monkey speech would obviously not sound the same as human speech, but that nevertheless, 'a monkey vocal tract would be able to produce clearly intelligible speech.'

For now, there are no firm answers in this debate. Perhaps monkeys are very far down the evolutionary path to language, but just not far enough. Perhaps 'monkey speech' would sound very strange, but still be intelligible. Perhaps the Simian Tongue will one day be decipherable and found to be the most extraordinary language of all.

The Power of Words

Bringing together the threads of how and why we evolved to speak, read and write teaches us something profound about language: it shapes our thoughts and changes the way we perceive reality. Consider how we experience colour. The Dani tribe of Papua New Guinea recognise only two colours: 'mili' for cool dark colours such as blue, green and black, and 'mola' for warm light colours such as red, yellow and white. Russian speakers lack a word for blue: instead shades of light blue are 'goluboy' and dark blue, 'siniy'. Pink is just a light shade of red, yet it's treated as its own colour in English. Australian aboriginals have no words for left and right: their language uses only north, south, east and west. Some languages, such as German and Spanish, assign every noun with a gender; the word 'bridge' is feminine in German, masculine in Spanish. When asked to describe a bridge, German speakers are more likely to use conventionally feminine words such as 'beautiful' and 'elegant', whereas Spanish speakers are more likely to use conventionally masculine words such as 'strong' and 'sturdy'. The examples are endless. One is tempted to speculate that with 7,000 languages come 7,000 cognitive realities.

For the time being, the language that dominates the human mind is unequivocally American English. Nearly 400 million people speak it as a first language and more than 1.6 billion as a second. It serves as the official language of fifty-nine countries and is the universal language of business, diplomacy, the Internet and science. The literary critic Jonathan Arac has remarked, 'English in culture, like the dollar in economics, serves as the medium through which knowledge may be translated from the local to the global.'[17]

Historically, English spread due to the expansion of the British Empire between the seventeenth and nineteenth centuries, and the emergence of the United States of America as the world's superpower in the twentieth century. A dark history of conquest and colonisation accompanied this expansion, leading many to view English as a form of linguistic imperialism. To this day, English continues to erode the native languages of postcolonial settings such as India, Pakistan, Uganda and Zimbabwe.

Like all empires, the English language may eventually start to decline. If it does, Chinese will probably take its place, a wondrous language which appeared over 3,000 years ago and is spoken by 1.1 billion people in the world today. Business experts have pointed out that Chinese is already becoming the language of choice in global markets. And as China's economy continues to outpace America's, Chinese could soon supplant English as the most widely used Internet language.

The Chinese language has a fascinating history. Belonging to the Sino-Tibetan family of languages, it is the oldest written language in the world, with a dictionary containing some 40,000 characters. Symbols that bear a striking resemblance to these characters have been found inscribed in 8,600-year-old tortoise shells at Jiahu in the Henan Province of western China; they appear to anticipate the Chinese characters for 'eye', 'window' and the numerals 1, 2, 8, 10 and 20. Latin didn't appear for another thousand years, well into the seventh century BC. As the Chinese language evolved, it transformed into a collection of languages that spread far and wide – to Malaysia, Singapore and Indonesia. The main Chinese language is Mandarin; others such as Yue Chinese (Cantonese), Min Chinese and Hakka Chinese are now among hundreds of mutually unintelligible varieties.

Would the fall of English and the rise of Chinese or another language shape our thoughts and change the way we perceive reality? Almost certainly, writes linguist Nicholas Ostler, who chairs the Foundation for Endangered Languages (the world loses a language every two weeks). Ostler believes that by 2050 English will cease to be the world's *lingua franca*, changing much more than how we speak to one another:

> If other centres emerge, the result may be more mixed, asserting local Islamic, Buddhist or Hindu traditions. Evolving translation technologies may make languages largely interchangeable, pushing national cultures into the background. Whatever, there will be no special deference to the current English-speaking tradition.[18]

Whatever happens in the future, we first have to remember that language evolved for one reason – to unite us. By creating abstract mental terms for what a hunter-gatherer 'sees', 'wants', and 'knows', language allows them to build a relationship with others and form a social group. For most of human history this was done by small pockets of people that were totally isolated from one another. Neighbouring tribes were thus seen as a threat to a group's social identity, and a competitor for scarce resources. Cue social conflict and a blood-drenched history of human confrontation.

But like groups, languages are living creatures that adapt to the times. As human populations grew, mixed and coalesced into ever-expanding cities, languages branched into different dialects and families. Some of these would have been understood by multiple sets of speakers – they had what linguistics call mutual intelligibility. But over time, as groups spread to the point where languages collided, and trade and religion required a way for humans to communicate with each other, mutual intelligibility gave way to translation.

Translation dates back to earlier than 1000 BC; in a written document from the Chinese Zhou dynasty, the imperial scholar Jia Gongyan declared, 'translation is to replace one written language with another without changing the meaning for mutual understanding.' Translation helped make the world a global village, and abstract terms for what a human 'sees', 'wants' and 'knows' can now be decoded in order to build relationships *between* social groups. This act of transference ushered in international cooperation and the rapid advancement of ideas and social progress.

Language matters. Human minds are organs that use words to connect, sharing our experiences and transmitting them across generations in the form of ancient texts, books and, increasingly, social media. We may never fully understand how language arose – how ape-like communication gradually transformed into the poetic prose of Vladimir Nabokov or the graceful Arabic of *One*

Thousand and One Nights. But 'how', in this case, is less important than 'what'. Because more than anything, language represents the culture of a particular group by codifying their values, beliefs and customs. It's a social phenomenon; indeed its social function is its evolutionary *raison d'être*. But as powerful as language is, it would not have been possible without the brain change that knitted all our thoughts together – a phenomenon that some believe evolved only a few thousand years ago: the illusion that is consciousness.

7

The Illusion of Consciousness

What is it like to be you? To wake up every morning, look at yourself in the mirror, and go about your daily life? What is it like to think all the things you think, to feel all the things you feel? It must be at least somewhat different from being me: whoever you are, you have your own history, your own experiences, your own memories, thoughts and desires. Your own life. Your own sense of *being you*.

And so we come to arguably the biggest mystery of the human brain: consciousness – our subjective experience of the world and all its perceptual contents, including sights, sounds, thoughts and sensations. A private inner universe that utterly disappears in states such as general anaesthesia or dreamless sleep. And something so mysterious that we still find it notoriously difficult to understand or even define.

Many have tried. In his famous 1974 essay, 'What is it Like To Be a Bat?', the American philosopher Thomas Nagel asks us to imagine changing places with a bat. His interest wasn't in bats but in making the point that an organism can only be considered conscious 'if and only if there is something that it is like to *be* that organism – something that it is like *for* the organism.'[1] We could call this the subjective experience of being a bat; a state of being that is comparable to the bat's.

Let's take Nagel up on his challenge and imagine being a bat. A bat's experience must be starkly different from our own. Most use echolocation to navigate and find food, releasing sound waves from their mouths or noses that bounce off objects and return to their ears, informing them of the object's shape, size and location. Some bats glide through the air releasing slow and steady pulses of sound, which then rapidly speed up when they swoop down on their prey. Others calculate their speed relative to their prey using the Doppler effect (the change in sound frequency that happens when the source and/or the receiver are moving; the same reason an ambulance siren sounds differently as it passes). Being a bat, I imagine, would be to live in a shadowy, kaleidoscopic world of sound, instinct and twilight flight.

But is this really what it would be like, or have I simply tried to imagine that *I* am a bat? If there is in fact something that it is like to be a bat, is it merely a sense of bat subjectivity, or something more? It's hard to say.

In the 1990s, the Australian philosopher David Chalmers took things further, proposing a hypothetical entity called the 'philosophical zombie': an exact, atom-for-atom duplicate of a human, indistinguishable from a real person in all its behaviour, only with no conscious experience whatsoever. Spooky, right? I envisage such a being to be a bit like Patrick Bateman, the protagonist villain of Bret Easton Ellis's novel *American Psycho*, who at one point in the story reveals,

> There is an idea of a Patrick Bateman, some kind of abstraction, but there is no real me, only an entity, something illusory, and though I can hide my cold gaze and you can shake my hand and feel flesh gripping yours and maybe you can even sense our lifestyles are probably comparable: I simply am not there.

Bateman is terrifying not for what his mind contains but for what it lacks. And here's the point: if philosophical zombies are possible, Chalmers argued, it follows that conscious states might not be entirely connected to brain states – that there is something more to conscious life than neurons firing inside the brain.

If bats and zombies aren't your thing, consider Mary the colour scientist. Mary specialises in the neurophysiology of colour vision, and thus knows everything there is to know about colour perception. She knows precisely how different wavelengths of light impinge on the retina and stimulate photoreceptors. She knows how they convert light into signals that are sent up the optic nerve to the primary visual cortex in the brain. And she knows all the cellular and molecular details of how the visual system eventually produces the experience of blue, green, red and so on.

But Mary has spent her entire life in a black and white room. She has never actually seen any colours; she has learnt about them and the world through black and white books and television programmes. One day, Mary escapes her monochrome prison and sees a brilliant blue sky for the first time. What changes? Does Mary learn something new upon seeing blue for the first time? Or is she unsurprised, since she already knows everything there is to know about how the brain processes blue in advance? If you think Mary learns something fundamentally new about the colour blue, you may consequently believe that physical facts about the world are not all there is to know.

Science still has no answer to these mind-bending thought experiments, but they are valuable because they encourage philosophers and neuroscientists to work together, to reconsider previous models and build a scientific framework for new accounts of how the brain gives rise to conscious thought. Most are essentially updated versions of the great philosopher René Descartes' mind–body dualism. In *Meditations on First Philosophy* (1637), Descartes concluded that the mind was immaterial, something totally distinct from the physical properties of the brain. Consciousness, from this view, wasn't so far removed from the Judeo-Christian notion of a soul, and indeed Descartes was strongly influenced by the Augustinian tradition of dividing soul and body. The resulting 'Cartesian' biology came to dominate thinking until 1949, when the British philosopher Gilbert Ryle ridiculed dualism as 'the dogma of the ghost in the machine'.

Such thought experiments, however, can be misleading. Some scholars have pointed out that it is in fact tremendously difficult to imagine knowing *everything* there is to know – about colour, for instance. In consequence, we may be tying ourselves up in philosophical knots, mistaking what is merely a failure of imagination for genuine insight.

If this all sounds terribly confusing, that's because it is. And it will remain so until we solve what's called the 'hard problem' of consciousness: namely, why are *any* physical processes in the brain accompanied by conscious experience? If the brain is ultimately just a collection of molecules shuttling around inside the skull – the same molecules that comprise earth, rock and stars – why do we think and feel anything at all? Why does our extraordinary mind spring from soggy grey matter to begin with? It's a problem that's been with us for centuries, as opposed to the 'easy problem' of consciousness, i.e. explaining how the brain works. Examples of easy problems include the biology of neurons, the mechanisms of attention and the control of behaviour – practical problems that relate to our experience of the world and that are not as deeply mysterious as the hard problem. Problems we know we can solve, in other words.

Some neuroscientists believe we will never solve the hard problem. Just as a goldfish will never be able to read a newspaper or write a sonnet, *Homo sapiens*, these scholars argue, are cognitively closed to such knowledge. It is a great but impenetrable mystery. The psychologist Steven Pinker calls the hard problem 'the ultimate tease . . . forever beyond our conceptual grasp.' Echoing the view that consciousness remains outside the limits of human comprehension, one of the best entries in Ambrose Bierce's *The Devil's Dictionary* is the following:

> **Mind**, *n.* A mysterious form of matter secreted by the brain. Its chief activity consists in the endeavor to ascertain its own nature,

the futility of the attempt being due to the fact that it has nothing but itself to know itself with.

Others believe that if we just keep solving the easy problems, the hard problem will disappear. By locating and understanding what we call the neural correlates of consciousness (NCC) – neural mechanisms that researchers say are responsible for consciousness, typically gleaned using brain scans or neurosurgery to compare conscious and unconscious states – we will march ever closer to solving the mystery, until one day there is nothing left to solve. Defining an NCC starts as a process of elimination: the spinal cord and cerebellum can be ruled out, for instance, because if both are lost to stroke or trauma, nothing happens to the victim's consciousness. They still perceive and experience their surroundings as they did before. The best candidates for NCC (so far) are a subset of neurons in a posterior hot zone of the brain that comprises the parietal, occipital and temporal lobes of the cerebral cortex. When the posterior hot zone is electrically stimulated, as it sometimes is during surgery for brain tumours, a person will report experiencing a menagerie of thoughts, memories, sensations, visual and auditory hallucinations, and an eerie feeling of surrealism or familiarity. So if the consciousness illusion is located anywhere, it might be in this mysterious region of the posterior cortex.

Still others believe that finding the neural correlates of consciousness will help us reveal the anatomy of consciousness, that is, its biological constituents in the brain, but that this will still not solve the hard problem because not all NCC will necessarily be a part of consciousness in the first place.[2] It would be like scrutinising the inside of a computer and declaring, 'Aha! This is where the Internet lives.'

There is another possibility, one I find the most compelling. There is no hard problem – because there is no such thing as consciousness in the first place. We have been misled, partly by the intriguing but ultimately false philosophical traditions of our forebears – most of whom were theologians first and scientists second – but mainly by the language we use to talk about consciousness. Whenever we say something 'enters consciousness', or 'shifts conscious experience', we are resurrecting Descartes' mind–body dualism, creating a new kind of Ryle's 'ghost in the machine'. We fail to realise that the physical activity of our brain *equals* our conscious experience of the world. Any separation is to commit a category error: a fallacy that arises when we assign something with a quality that belongs to a different category. It would be a category error, for example, to ask about the lyrics of flowers, even though it might sound deep and mysterious, because flowers are categorised by genus, species, and often variety, but never by lyrics. Similarly, it would be a category error to visit the churches and galleries of Florence only to ask, 'Where is the city?'.

The mistake we make when pondering consciousness is to place feelings and experiences into some mystical, nebulous category that is separate from the brain itself. But if I put my brain in a blender and ground it into a pulpy mush, all my conscious thoughts and feelings, all my memories and perceptions, would instantly vanish. To claim consciousness is 'one's experience' is a tautology: you are your experiences, and your experiences are you.

The truth is that we still, despite all our enlightened science and empirical reasoning, struggle to simply accept the reality of materialism – the reality that there is only matter and energy in the universe, bound together by the scientific laws of nature. There are no non-physical minds or immaterial thoughts. Conscious states *are* brain states. As Nick Chater declares in his wonderful book *The Mind is Flat*,

> There are no conscious thoughts and unconscious thoughts, and there are certainly no thoughts slipping in and out of consciousness. There is just one type of thought, and each such thought has two aspects: a conscious read-out, and unconscious processing generating the read-out. And we can have no more conscious

access to these brain processes than we can have conscious awareness of the chemistry of digestion or the biophysics of our muscles.[3]

No doubt, this view gives short shrift to hypothetical bats and colour scientists. But good riddance, I say. Dualism is a philosophy that has plagued neuroscience for too long. Supporters of Nagel et al. might argue that even when we fully understand the neurobiology of consciousness, we will still be missing something essential. But what, exactly? No matter how you slice it, every conceivable answer relies upon mysticism and the denial of modern neuroscience.

We've been here before. For centuries, the theory of a 'vital force' (or *élan vital*) was proffered to explain how living things could arise out of non-living matter. The fire-like element phlogiston was once needed to explain oxygen; caloric fluid was advanced to explain heat. The illusion of consciousness is almost as absurd as the creationist notion of 'irreducible complexity': the belief that the molecular complexity of things like the bacterial flagellum is unassailable proof that it could not have arisen by a gradual step-by-step evolutionary process; God must be responsible.

Some people baulk at materialism because they fear it will denude life of its meaning. If human minds are just bags of chemicals with no mental depths to plumb, if all our actions are just the result of the electrochemical activity of neurons performing intelligent computations, what is the point of life? Here again the crisis stems from a misuse of language. 'Just' is not appropriate because these processes are extraordinary and many of them are beyond our current conception.

The reason we find materialism hard to swallow is because the words we use to describe brain activity are different to those we use to describe feelings. No one has trouble accepting that rain or snow are just other ways of discussing annual precipitation, but try telling someone that anxiety and happiness are just other ways of discussing serotonergic neuronal activity. Both types of language are part of the same equation, yet we focus on only one side.[4]

None of this is meant to remove consciousness from our evolutionary journey, any more than my emphasis on nurture removes the salience of nature. It is meant to topple our preconceived notions of consciousness in order to focus on what really matters: why consciousness *seems* the way that it does. To answer this question, we must explore how – and why – the illusion of consciousness evolved in the first place.

How Illusions Evolve

An illusion is not something that doesn't exist; it is something that is not what it seems to be. When I wake in the middle of the night and perceive the washing on my clotheshorse as if it were a small child, I am – thankfully – having an illusion. The socks and T-shirts really do exist, and could (with a little artistic license) be taken to resemble a child. But it's not a child. Just as the sun isn't orbiting a flat earth, my mind has misinterpreted the facts. The same thing happens when we look at a missing-shapes illusion. The black shapes really do exist, but there is no white triangle or white sphere. Your brain has invented them.

With respect to consciousness, our experience exists but the notion of consciousness does not. It may seem as though our experiences are part of something larger – some grand cohesive narrative, a stream of consciousness, a movie-in-the-brain – yet this is no more real than the child I sometimes see in my clothes-horse. It is 'a magic show that you stage for yourself inside your own head,' says the psychologist Nicholas Humphrey.

Still, even magic shows need explaining. Did the illusion evolve for a particular purpose or is it a useless by-product: an evolutionary vestige like the human tailbone and goose bumps? In my view, whatever purpose the consciousness illusion might serve, it is principally no more than the emergent phenomena of brain activity – no more mysterious than the fact that sand forms dunes and birds fly in flocks.

Consider frogs. Their eyes do not tell their brains all the things that human eyes tell the human brain. Unlike humans, frogs do not construct a picture of the world in their heads; they see only what they need to see, namely moving edges, a little light, and insects. Inside a frog's eye, in fact, is a network of nerves that act as an insect detector, which activates only when small moving objects enter the frog's visual field, telling it to release its tongue to catch its prey. Remarkably, a frog can starve to death surrounded by dead flies – because if the flies don't move, the frog doesn't see them. Poor frogs, you might think. Yet we may be more like them than we realise, for we too have only a partial perception of reality.

In the 1950s, in an effort to treat people with severe epilepsy, some patients received a *callosotomy*, a surgical procedure that severs the corpus callosum: the nerves dividing the brain into left and right hemispheres. The surgery appeared to work, alleviating the patients' seizures without producing any discernible side effects. These so-called split-brain patients allowed scientists to investigate the relative contributions of the 'left-brain' versus the 'right-brain'.[5] While it is a myth that some people use one side of their brain more than the other – that left-brained people are more analytical whereas right-brained people are more creative – the hemispheres do possess a high degree of functional independence. The left side of the brain controls the right side of the body, for instance, and vice versa. Anything that falls in the left visual field of each eye is projected to the right hemisphere, and vice versa. And the left hemisphere is mainly, though not exclusively, in charge of language. But the hemispheric independence of a split-brain patient also revealed something eerily puzzling about conscious experience.

In the experiment, a split-brain patient is asked to focus on a dot in the middle of a screen. Then a word – say, 'spoon' – is briefly flashed to the left of the dot, which only the patient's right hemisphere sees. When asked what was seen, the patient (using her language-centred left hemisphere) will say 'nothing'. Yet when the same patient is asked to pick up the 'unseen' object from a variety of objects using only her left hand (which is controlled by the right hemisphere), she will correctly select a spoon every time. If the patient is then asked what she now holds in her left hand, she will not be able to say what the object is until her left hemisphere is also allowed to see it. When everything is revealed, and the person doing the experiment asks her why she picked the spoon, she will either claim not to know or invent a story to fill in the gaps. This is very weird. It suggests that parts of the brain operate independently of each other, creating the impression of something being unconscious. It suggests that evolution has also hidden things from us.

Consider too the phenomenon of binocular rivalry. When one eye is presented with one image – a bus, say – and the other eye is presented with a different image – a tree, say – we do not experience a blending of the two images. Instead, the bus will be seen for a few seconds, and then the tree, and then the bus again, and so on, constantly flipping back and forth until we decide to look away. This is also pretty weird. Despite being completely awake and responsive to our surroundings, a complete overhaul of the contents of our conscious experience is occurring all the time.

We're not really sure why the mind evolved this way, but just as wavelengths of light that our eyes cannot see and frequencies of sound that our ears cannot hear exist, evolution has clearly designed the brain to detect only particular features of the world around us. Like a glorified insect detector, the brain excludes unimportant details and zooms in on what really matters.

So where did the consciousness illusion come from? When and how did it evolve? For centuries these questions remained within the purview of theology and philosophy. Today there is a menagerie of evolutionary theories dedicated to this question. Some argue that consciousness arose as soon as single-celled life emerged, some 4 billion years ago. Others think that consciousness requires a nervous system, the first of which dates back to a wormlike organism about 550 to 600 million years ago, during the so-called Cambrian explosion. Still others, including the neurobiologist Bernard Baars, link consciousness to the arrival of the mammalian brain around 200 million years ago. The Oxford University neuroscientist Susan Greenfield thinks that consciousness evolved gradually as the brain size of hominids increased.[6] The Cambridge University neuroscientist Nicholas Humphrey thinks that the consciousness of *Homo sapiens* may have been the first kind of consciousness to evolve.[7]

To reach the truth, we need to consider three viewpoints. The first is that only humans experience the illusion of consciousness. If you think that language and other uniquely human capacities are necessary for the illusion, you may agree with this viewpoint. I will add, though, that the brains of our ape cousins are organised in a remarkably similar way to our own, and research shows that chimpanzees are self-aware and may even be able to comprehend the minds of others. So it does look as though they almost certainly experience some form of the consciousness illusion. The second is that all species experience the illusion. An extreme version of this view is sometimes called panpsychism – the idea that everything from cosmic dust to a blade of grass has some form of consciousness. However, even if such consciousness does emerge from the interaction of subatomic particles, that still leaves us with the riddle of how these 'micro-consciousnesses' combine to create human experience. The third is that consciousness can be split into sensory consciousness – defined as our awareness of sensations in the present moment (the smell of fresh coffee, the blueness of the sky) – and higher-order consciousness – defined as our *self*-awareness, our thoughts about our thoughts and our ability to contemplate our fundamental nature and purpose. If true, this might mean that we have an on/off switch for consciousness, which some scientists believe exists deep in the brain. Nevertheless, by breaking up consciousness into parts including attention, sensation, introspection and so on, one could argue that we are no more explaining the illusion than explaining the purpose of a birthday cake by discussing flour types, flavour combinations and baking time.

Consciousness is likely to be a relatively modern invention. The brain continually changes and develops in response to input, with culture perhaps exerting the greatest influence of all. Given our capacity for symbolic thought, the illusion probably co-evolved with the appearance of specialised social skills, language and creative thinking. Today, a new theory called the Attention Schema Theory (AST) suggests that consciousness arose at some unknown point between 200,000 and 2 million years ago to deal with the continuous excess of information flowing into our minds.[8] Human species during this time period ranged from *Homo habilis* to *Homo erectus* to *Homo heidelbergenis* to *Homo neanderthalensis* to *Homo floresiensis* to *Homo sapiens* (to name but a few). And each possessed a different brain size which, as we have seen, gave rise to different cognition and behaviour. But there's a unifying factor about these early humans that helps explain where and how consciousness arose: their lives were a relentless test of where to focus their attention. In their struggle for survival, they needed to process a continuous excess of information. And so, by a strange form of neuronal competition, the theory goes, early human neurons had to select, moment-to-moment, which signals they were willing to convert into conscious awareness and which they were not. In essence the brain chose what it needed to know.

In the AST's evolutionary story, attention is the essential prerequisite for consciousness. For example, when a member of the *Homo erectus* species crept along the lake shore to hunt an antelope, her brain would first have had to suppress any competing signals for her attention, such as a bird or the wind rustling the grass. Then, as she got closer to the antelope, her attention became more focused; as far as her brain was concerned, it was now just her, the antelope, and her attention to the antelope. By constructing this level of attention, the brain concluded that such attention must belong to a separate, non-physical type of awareness. And with that, the illusion of consciousness was born. This theory fits nicely with our observations regarding optical illusions and split-brain patients, and also explains how the hardware for conscious thought could have emerged gradually over millions of years.

Then, enter *Homo sapiens*, fitted with a brain capable of igniting a social, linguistic and cultural explosion, roughly 60,000 to 40,000 years ago. As the brain expanded, the amount of information it could store and manage increased, leading to the phenomenon of cultural evolution. An important consequence of cultural evolution is a set of shared attitudes, values, beliefs and behaviours. Our brains have chosen what they need to know based on our social milieu. The consciousness illusion is not fixed. It evolved in tandem with the society that we created. It is our social and cultural experience of the world. The consciousness illusion is whatever we want it to be.

The Self Illusion

I opened this chapter with a question. Did you answer it? Could you answer it? Whatever you thought, I suspect a part of you considered it a reasonable question. Our everyday language is saturated, indeed obsessed, with our sense of self. Though our opinions, desires, hopes and fantasies may change over time, we return always to ourselves. As James Joyce wrote in *Ulysses*, 'Every life is in many days, day after day. We walk through ourselves, meeting robbers, ghosts, giants, old men, young men, wives, widows, brothers-in-love, but always meeting ourselves.'[9]

But this way of thinking about ourselves is problematic. By talking about the brain as something that 'I' possess, I separate myself from it, thereby creating an illusory being, an everlasting author, who does not in fact exist. The Buddha understood this. In his sixth-century-BC doctrine of *anatta* he taught that the self is a changing, impermanent multitude of discrete parts: an individual's identity, he said, is selfless. Building on this idea, the Oxford philosopher Derek Parfit (1942–2017) advanced two theories of personal identity: the ego theory and the bundle theory. Ego theorists believe there is a single, continuous self controlling our identity. Bundle theorists, on the other hand, believe there is no one self but a bundle, or collection of selves that ultimately defines us. Buddha was perhaps the first bundle theorist.

And certainly not the last. In *A Treatise of Human Nature* (1739), David Hume argued that the self is merely a bundle of different perceptions. It only feels like a continuous entity because memory seems to tie all our experiences together. But this is just another illusion. There is nothing supplementary to the feeling of continuity, no divine witness of the world and your earthly experiences. Hume writes:

> For my part, when I enter most intimately into what I call *myself*, I always stumble on some particular perception or other, of heat or cold, light or shade, love or hatred, pain or pleasure. I never can catch *myself* at any time without a perception, and never can observe any thing but the perception.[10]

Like the Buddha, Hume discovered a fundamental truth about human experience: there is no experiencer. This was seen as heresy to ego theorists of the time. As Hume's contemporary, the Scottish philosopher Thomas Reid, retorted: 'I am not thought, I am not action, I am not feeling: I am something that thinks, and acts, and suffers.'[11] But ponder what Reid is saying here. He is 'not thought', but 'something that thinks'. What is that something, exactly? It is millions of interconnected neurons firing in the brain. It is a physical substance giving rise to behaviours, memories, and perceptions. It *is* thought. He then says he is 'not action', but something that 'acts'. Well, okay, but actions and decisions occur due to the biological laws of living systems. The fact that we don't yet fully understand those laws does not legitimise an illusion

More importantly, the concept of a unified self is bad science because it is unfalsifiable, that is, there is no experimental test to disprove it. A theory is only considered good science when it is falsifiable, meaning it is possible to devise an experiment that disproves the idea in question. If I hypothesise that all apples are sour, for instance, a quick test in the form of me eating a sweet apple will falsify my hypothesis. It might sound like a silly hypothesis, but the key point is that it is *testable*, because it is based in reality. Horoscopes are good examples of something that isn't falsifiable. No one can disprove that a Gemini will have a dispute with a close relative, or whether a Capricorn wearing something light blue will be lucky between 11.00 a.m. and 1.15 p.m. The statements are too vague. Even if those predictions didn't come true, that wouldn't make them false because they still could have happened. The same is true with the self. It is too vague to be falsifiable.

Ask any neuroscientist today and she will probably agree that the illusion of self, while certainly being a subjective experience, is one that simply does not exist in nature. Just as cosmology no longer needs a God to explain the universe, neuroscience no longer needs a unified self to explain the mind. Like it or not, when you contemplate the existence of a self you discover that it is a mirage – just another form of mind–body dualism.

Ask yourself, am I the same 'me' as I was a moment ago? When you really think about it, the answer has to be no. After all, are you the same 'you' you were ten years ago? It's doubtful. I for one am almost nothing like the person I was ten years ago; I've grown up, I've changed. Something we all do in one form or another. As the British philosopher Galen Strawson observed, 'Many mental selves exist, one at a time and one after another, like pearls on a string.'[12]

The reason it feels like we have a continuing inner self is because we each possess what neuroscientists call the phenomenal self-model (PSM). This, according to its German pioneer Thomas Metzinger, is 'a distinct and coherent pattern of neural activity that allows you to integrate parts of the world into an inner image of yourself as a whole.'[13] In the brain, the PSM is thought to be located mainly in the prefrontal cortex, and some believe it is impaired in schizophrenia and extreme narcissism. To be clear: the PSM is not the self. As Metzinger notes, 'no such things as selves exist in the world: nobody ever *was* or *had* a self.' The PSM is merely the neural circuitry required for our feeling of ownership and agency – the ability to perceive your limbs as *your* limbs, for instance.

Today, we ceaselessly create new selves via social media platforms such as Facebook, Twitter and Instagram. These selves are often very different from the selves we manifest in face-to-face encounters: they each possess a distinct narrative, a unique voice and, usually, an idealised self-image. Moreover, each self can affect the other, producing a kaleidoscope of psychological complexity.

A startling consequence of multiple selves is that each can be a product of the social setting it finds itself in. Scholars in the new field of discursive psychology, a branch of science that investigates how the self is socially constructed, are particularly interested in how people's pronouns affect their sense of self. According to these scholars, the self is a 'continuous production' built from words and culture. Until recently, the first-person pronouns 'I', 'me' and 'mine' predominated in Western society. Now, however, dozens of pronouns are used to express various identities: 'ze', 'ey', 'hir', 'xe', 'hen', 've', 'ne', 'per', 'thon' and more. Sticklers for grammar view this as an assault on the English language. Sticklers for tradition view it as a slippery slope to government-mandated speech codes. Yet far from being new-age gobbledygook or an omen of tyranny, novel pronouns in fact have a crucial social function. They reveal your personality, reflect important differences among groups and help knit communities together.

The construction of the self also appears to be a phenomenon in the wider animal kingdom. For decades, scientists have been searching for signs of self-awareness in animals. The most widely used test, called the mirror test, works by applying a spot of odourless dye to an animal's forehead and then seeing if it tries to scrub off the mark when placed in front of a mirror. Animals that pass this test include chimps, bonobos, orangutans, Asian elephants, Eurasian magpies, whales and dolphins (which inspect rather than scrub off the mark). Gorillas and rhesus macaques usually fail the test. Humans recognise themselves in a mirror by the age of three.

Recent findings have challenged whether the mirror test measures self-awareness, or the ability to *learn* self-awareness. In 2017, Liangtang Chang, a researcher at the Shanghai Institutes for Biological Sciences, China, and his colleagues trained rhesus macaques to recognise themselves in a mirror.[14] To do this, they gave the monkeys a food reward for touching a red laser dot projected onto a surface in the monkeys' vicinity. At first the laser was shone directly in front of the monkeys, where they could easily spot it. Then, it was shone on an area that could only be seen through a mirror. Following several weeks of such training, the researchers started moving the laser dot to the monkeys' face. And *voilà*! They suddenly realised that the red dot on the face in the mirror was pointing to their own. They had learned self-awareness.

After their training, the monkeys were filmed exploring their newly discovered skill, inspecting their facial hair and fingertips, investigating previously unseen body parts and playfully flashing their genitals. The video footage leaves little doubt: they are completely hooked on themselves, enthralled by their own existence. 'Clever studies like the one by Chang [and colleagues],' wrote a group of fellow researchers, 'help expose our preconceptions about ourselves and point the way toward a deeper understanding of the way our brains, and the brains of other animals, construct reality and our place within it.'[15] That deeper understanding is twofold: it suggests that other species have at least one sense of self, and that the social construction of the self means that self-awareness can be taught.

There are good reasons why our minds evolved this way. The brain is constantly building models of the world around us. For early humans, life in tight-knit communities meant they probably only needed to construct a few mental selves to survive and reproduce. Gradually, as social groups grew larger and more complex, each individual member of *Homo sapiens* needed to construct multiple selves in order to adapt to the new social system.

The notion of a unified self is not only an illusion but a dangerous one. By rejecting the diversity of ways it is possible to live in the world, the unitary self inhibits mutual respect and social justice. Importantly, it's now well-established that our attachment to the unified self can cause bullying, selfishness, and greed, which sometimes tips into narcissistic personality disorder (NPD): a mental health condition characterised by arrogant thinking, lack of empathy and an inflated sense of importance.[16] Individual rights are vital for a free society, no doubt; but to become a moral society we must be more open-minded about the *community* of selves we each are.

Because the truth is, the self isn't one thing. It's a spectacular profusion of things – it's the social self, the solitary self, the loving self, the caring-for-the-environment self, the learning-something-new self. There is no owner, no core self. Nothing about us is centralised. We are each a multiplicity of ever-changing, reinvented, redefined selves. The French philosopher Jean-Jacques Rousseau vividly captured this astonishing revelation in his *Confessions*:

> I have very intense passions, and while I am in their grip my impetuousness is without equal: I know neither restraint, nor respect, nor fear, nor decorum; I am cynical, insolent, violent, bold: there is no shame that could stop me, nor danger that could frighten me: beyond the one object that occupies me, the universe is nothing to me. But all that lasts only a moment, and the moment that follows annihilates me.[17]

Free Will

The realisation that our brains evolved to conjure up illusions and multiple selves leaves a big question to be answered: are we free? Is what we do a choice freely made or the result of inevitable computations? The classic problem of free will is one of the most misunderstood in all of neuroscience. For centuries, theologians and natural philosophers have been at loggerheads over the nature and significance of free will. It's a debate that's filled every period of world history and been taken up by every intellectual figure since the dawn of our species.

Plato believed that free will is entirely based on our knowledge of good and evil: the person who strives for good in the form of wisdom, courage and temperance is truly free, while the person who abstains from such virtues and seeks pleasure and passion alone is a slave. It became known as ethical determinism – or 'moral liberty' in the Christian tradition – and has been discussed by philosophers of all persuasions. Another Greek giant, Aristotle, took things further by arguing that good choices lead to good habits, which in turn lead to good character. Like Plato, he believed that humans can freely choose between right and wrong, but he placed more emphasis on free will being something earned as a result of good character.

The idea that human choices may be partly or wholly determined by natural forces beyond our control was first put forward by the Stoics, a school of philosophy founded by Zeno of Citium, in Athens, in the early third century BC. Zeno and his disciples believed that everything in the universe and all human behaviour is causally determined by the laws of science. They still believed that our choices can be freely made – life presents us with options and we can clearly see that those options have their own consequences. We still therefore have a moral responsibility to make good choices.

The Stoics had their critics. Among them was the philosopher Alexander of Aphrodisias, who, in his work *On Fate*, argued against any earthly influence on human action. Shortly after this, monotheistic theologians including Saint Augustine of Hippo (354–430), Al-Ash'ari (874–936) and Maimonides (1138–1204) used religious dogma rather than reason to understand free will. It was the Enlightenment philosophers such as Thomas Hobbes (1588–1679), David Hume (1711–76) and Immanuel Kant (1724–1804) who reaffirmed that free will was written in the clockwork laws of nature.

Since then, modern conceptions of free will have focused on evolution and the inner workings of the brain. Just as the firing of neurons determines our thoughts, memories and dreams, so too do they govern our choices, behaviour and sense of free will. Nowhere is this more evident than in cases of ordinary people who become killers or rapists after suffering from brain tumours that fundamentally alter their brain chemistry. As unsettling as it may seem, these people no more choose their actions than a hurricane chooses to destroy a building. The same rubric applies to healthy people, as well. Were we to understand the brain better than we currently do, we could predict a person's future behaviour with astonishing accuracy.

This brings us to science's most up-to-date understanding of free will: you have free will, you just don't have conscious will. How do we know this? Well, as far back as the late 1980s, an American neuroscientist named Benjamin Libet conducted a series of experiments which revealed that we become consciously aware of a decision about half a second *after* the brain has directed the neural mechanisms responsible for the decision.[18] Think for a moment how strange this is. In our daily lives we experience freely made choices as being willed into existence: we consider the options before us, deliberate over which is best and, then, as with a muscle, we consciously 'flex' our choice into the world. Half a second is quite a long time. Wouldn't we notice such a delay?

The answer is no because, as we have seen, there is no distinct and separate self in the first place. It merely feels like we have made the decision when in fact the brain has already made it for us. You can begin to grasp this with an easy experiment. Take a moment to give yourself a simple choice. It can be anything – clapping your hands, lifting your leg, turning your head – but the choice must be made every twenty seconds. Repeat this for a few minutes, closely observing the experience and how it feels. Think of nothing else besides the choice and how it is being made. Can you tell how the choice is being made? Are you choosing when to choose, or is the choice just appearing in your consciousness? Does consciousness deliver the choice to you, or does the choice seem to arrive from nowhere? What made you choose to clap or not to clap? When you really think about it, you realise that 'you' – the self – had absolutely nothing to do with it.[19]

Some researchers think that half a second is underestimating the time it takes for our brains to make decisions. In 2008, a group at the Max Planck Institute for Human Cognitive and Brain Sciences in Leipzig, Germany, conducted brain scans on fourteen people while they performed a decision-making task.[20] They were asked simply to press a button with their left or right hand whenever they felt like it. When the team analysed the data, they detected a brain signal more than *seven seconds* before the volunteers claimed to have made their decision. Seven seconds is a lifetime to the brain. To put that in context, it took the world's most powerful supercomputer forty minutes to simulate just one second of brain activity.[21]

For the brain, the events taking place in that gulf of time are practically infinite. What's happening in those dark seven seconds is anyone's guess. The only thing the researchers spotted was a shift in brain activity from the frontopolar cortex (a region linked to planning) to the parietal cortex (a region linked to movement), which isn't a great deal to go on. But as the study's lead author John Dylan Haynes comments, the implications are clear: 'there's not much space for free will to operate.'

As with consciousness, the illusion of free will is a result of evolution. But how and why would evolution generate something so complex and deceptive? The answer to 'how' is easy: as another product of the brain, the free will illusion consists of the billions of neurons that evolution has built for us. No single neuron or brain region has free will, of course; not even the famous neurosurgeon Henry Marsh could find a piece of brain tissue that knows or cares who we are. But like much in the brain, the origin of free will is to be found in the cohesive whole and not its constituent parts.

The answer to 'why' isn't so straightforward, but a great deal of evidence suggests that while free will is an illusion created by our brain, we are better off believing we have it anyway. People who believe that they have control of their actions are happier, less stressed, less likely to steal, less likely to cheat in an exam, more creative, more humble, more grateful, more committed to relationships, more likely to work hard and show up for work on time and more likely to give money to the homeless.[22]

Amazingly, the illusion of free will can be triggered in people. In 1999 the psychologists Daniel Wegner and Thalia Wheatley performed experiments based on the concept of the Ouija board.[23] The only spirit being contacted in this version, however, was the everyday illusion of human agency. In the experiment, two participants were asked to move a board resting on top of a computer mouse, with the mouse moving a cursor over various pictures on a computer screen. One participant was then asked to force the other to land on a particular picture, a little like cheating on a Ouija board. Each participant was wearing headphones, through which they would hear the name of the picture at different time intervals: either slightly before or slightly after they are forced to move the cursor.

Just as on a Ouija board, it's obvious when the other participant is forcing the move. However, if the name of the object is played a second before the participant is forced to move the cursor, she or he will claim to have done it themselves. It's that easy to be fooled. Wegner calls this the mind's best trick. Despite walking around with the day-to-day certainty that we consciously control our behaviour, it just isn't so. The feeling of free will, he says, merely 'arises from interpreting one's thought as the cause of the act.'

Viewed in this way, human choice is more delicate and mysterious than we ever imagined. Silently cocooned in an illusory exterior, hidden in depths that may forever lie beyond our grasp, our choices emerge like dark matter in the cosmos. They affect everything we hold dear: our laws, our politics and our sense of moral responsibility. The fact that human minds evolved over millions of years by chance does not diminish their value. Our lives, as William James declared, are 'not the dull rattling off of a chain that was forged innumerable ages ago.' Perhaps, then, the most precious gift we have inherited from our brain's evolutionary journey is that we are all born free.

8

Different Minds

Sally Hopkins always sensed her mind was different. At her thirteenth birthday party, she sat silently in the corner, meticulously counting the number of times the other children jumped on the trampoline. At her end-of-school party, while her friends drank and danced all night, she spent the evening drawing intricately patterned animals, wolves and owls among her favourites. And at her first job interview – to be a waitress – she became unduly distracted by a coffee stain on her interviewer's shirtsleeve, and pointed it out to him several times.

'Suffice it to say, I didn't get that job,' Sally told me when I met her for a walk on London's Hampstead Heath. 'Or many others, for that matter, even though I tell them that I'm autistic.' It was a cool summer's day in July 2021, and I'd reached out to Sally via a colleague who specialises in autism research. I'd read about autism and met many people with the condition, but I'd never truly considered what the phenomenon meant for brain evolution. It was a brain change like any other and so warranted further exploration.

The heath was full of people, desperate to enjoy the outdoors after more than a year of Covid-19 lockdowns. Despite the end of local restrictions, we decided to abide by social distancing. 'My entire life has felt like an exercise in social distancing,' Sally noted wryly, 'so this actually feels normal to me.'

Sally is a laid-back, perspicacious woman with dark hair, bright blue eyes and an easy smile. Growing up in south Norfolk as the only autistic child among four siblings, she struggled to comprehend the reason for her uniqueness – opting instead to retreat into herself, chasing imaginary friends through the cobbled streets and thatched cottages of the English countryside. 'I spent a lot of time alone, but it was still an idyllic childhood. I felt connected with nature, with its rhythms and patterns. I liked to focus on the geometry of flowers; I still do.' Today, Sally spends her days writing poetry, devouring books and savouring the 'thrill' of freshly ground coffee. Her unique skills and talents include a remarkable memory, a preternatural eye for detail and a deep reverence for the environment. She also struggles socially in large groups and is averse to change.

As we strolled up Parliament Hill, the London skyline gently unfurling below, I asked Sally what she thought of her autism now, given that, at twenty-five, she was mature enough to acknowledge and apprehend it fully. 'I think of it the same way I did as a child,' she says,

> as a difference that other people find hard to accept. But I also don't know what to think about it, and I sometimes find myself pondering the concept of neurodivergence and what it signifies in the world. It's as if I've found an anomaly in the flower's geometry. I need to know what it means.

Like many autism advocates, Sally feels compelled to educate people about the condition. She regularly attends support groups and scientific conferences, and recently went round door to door to tell people to read Naoki Higashida's *The Reason I Jump*, a book recounting the experiences of its thirteen-year-old non-verbal autistic author. Higashida's goal in writing the book was to open people's minds to the endless wonders of autistic cognition. Like Higashida and so many others, Sally wants awareness, not sympathy; insight, not fear.

'I like talking to people like you,' she says, admiring a butterfly resting on a blade of grass.

People who want to understand things, to glimpse their inner purpose. I think most people are contented with the status quo. They find it comforting because it doesn't challenge them in any meaningful way. But maybe we can change that for autism. Or at least try.

What is Autism?

All minds are different, but some are more different than others. This is certainly true for Kim Peak, an American savant who can read both pages of an open book at once, his left eye reading the left page, his right eye reading the right. It's true for Stephen Wiltshire, a British artist who can draw the cityscape of Rome in meticulous detail, entirely from memory. It's true for countless techies in Silicon Valley. And it's been true for some of the world's greatest minds including Albert Einsten, Henry Cavendish, Nikola Tesla and Lewis Carroll.

We label this extraordinary difference 'autism' (from the Greek 'autos', meaning 'self', and 'ism' implying an isolated self). It's described as a developmental disorder characterised by learning difficulties, aversion to human contact, impairments in communication and social interaction, and the presence of often obsessive, repetitive behaviours. It's four times more common in boys than girls, and exhibits such a range of symptoms that it's now more generally referred to as autism spectrum disorder (ASD). First described in the 1940s by the Austrian paediatrician Hans Asperger (a eugenicist and Nazi sympathiser who sanctioned race hygiene policies including forced sterilisation), autism has captured the public's interest in a way few other conditions have. Powerful myths such as the cold, unloving 'refrigerator mother' were blamed for autism, and all manner of behavioural therapy has been called upon to treat it. Maybe – just maybe – the narrative goes, there is a way to release the 'normal' child locked inside.

Yet for all the frightening canards and scientific intrigue, few understand what autism is or why it exists. For decades, scientists have looked for signs of autism in the brain – for lesions, scars or tumours – all without success. Under an MRI scanner, the autistic brain shows some regions that are smaller than usual, but others that are larger. The cerebellum (the brain's movement centre) is 20 per cent smaller in some autistics, explaining why some experience problems with balance. An autistic person's visual cortex responds less strongly to faces than it does to objects and buildings, explaining why many people with autism find it hard to make eye contact. Patterns of communication between brain regions also differ in autistic people: those with high-functioning autism (referring to people with mild symptoms) have greater communication between some regions, but less between others. Under the microscope, the detailed structure of nerve cells shows the odd difference here and there. Some studies have found that neurons made from the stem cells of people with autism grow faster and develop more complex branches than their counterparts taken from cells of people without autism. Others have shown that the brain fails to prune synapses during childhood and adolescence, leaving a surplus of synapses that can overload the brain with information.

However, autistic brains also have distinct advantages. The autistic brain is on average 3 per cent larger than those without autism; in some cases it has been shown to be 15 per cent larger. Larger brains provide an evolutionary advantage, as we saw with the rise of *Homo sapiens* in the opening chapter. Larger-brained mammals are typically more intelligent, and for autistics the extra brain volume increases the thickness of the cerebral cortex (the brain's outer layer), creating a powerful memory system which many autistic people say accounts for their exceptional ability to recall visual details.

Genetically, autism is a deep mystery. Since the 1970s, researchers have known there is a genetic contribution, but no 'autism gene' has ever been found. We know today that some sixty-five genes are strongly associated with autism – those who carry any of them are more likely to develop autism – and that more than 200 other genes are weakly associated with it. These genes have subtle effects and work in concert. Their function is unclear, but many are believed to help neurons talk to one another or control how other genes are switched on and off. A gene called *neurexin*, for instance, helps neurons connect at the synapse and has mutated in a small percentage of autistics. Mutations have also been found in genes such as *neuroligan* and *SHANK3*, both of which are essential for the proper functioning of synapses. Though there are many other genes that help neurons connect, these findings lend credence to the idea that autism is ultimately a genetically induced synaptic difference.

The genetic picture becomes more mysterious when considering the role of copy number variations (CNVs): genetic changes involving duplications or deletions of vast stretches of DNA. Hundreds of CNVs have been linked to autism, but again they are rare, accounting for no more than 1 per cent of cases. Geneticists often speak of a 'many-to-one' relationship between genes and disease – there are many genes for Alzheimer's and Parkinson's, but usually only one set of symptoms. But for autism the relationship is 'many-to-many', because the many genes that can cause autism in some people lead to other diseases – such as epilepsy, ADHD and schizophrenia – in others. Today, researchers are looking for patterns among autism-related genes. But what such patterns would mean for autistics, no one yet knows.

In fact, how to interpret any of these findings remains unclear. Uta Frith, a neuroscientist who specialises in autism research at University College London, admits, 'We do not yet know the significance of any of these findings, what exactly they tell us about anatomical structure or physiological function . . . I cannot hide the fact that there is little specific to report about the causes of autism, or about the brain in autism.'[1]

There is of course another explanation. What if the way we've been thinking about autism is fundamentally wrong? What if autism is not a disease but rather a neurological difference? What if, given all the marvels of brain evolution we have seen, autism is simply another unexpected legacy of our evolving minds?

Ancient Savants

In 2004, an insightful archaeologist named Penny Spikins began to investigate the evolutionary origins of autism. Based at York University after conducting fieldwork in Patagonia and the Pennines, she started to wonder whether autism could have played a role in what made our species different.

'When you look back at the archaeological record,' she told me during a long conversation on Skype, 'you start to see transitions where there are increasingly finely made tools. The detail is astonishing. They are almost perfect. And that got me thinking . . . given that autism tends to create a greater attention to detail, what kind of contribution did individuals with autism bring to society?'

Spikins chose an auspicious moment to voice her radical idea. The rise of molecular genetics was transforming biology, forcing scientists to view the subtleties and complexities of autism in a completely new light. What's more, DNA dating was helping scientists understand the timeline of genes linked to autism in human evolution. And they were old. Very old. Some were older than multicellular life itself. In addition to being particularly sensitive to mutations, these ancient genes were notable for their role in programming the development of the body and the brain. Known in genetic parlance as 'hubs', they are often the starting point for larger gene networks to emerge over time. This makes them powerful contributors to genetic diversity.

'Autism-associated genes seem to be part of the, let's say, *evolvability* of the genome,' said Spikins.

> We know there must have been an element of positive selection for autism in the past. It must have bestowed certain advantages. And we know that some genes linked to autism are common in primates; that they're part of a shared ape heritage which predates the split that led to humans. There's even been some work on what you might call autistic traits in chimpanzees.

That work was done in 2016 by Kyoko Yoshida and his colleagues in Japan.[2] They noticed that one of their macaques was displaying some unusual behaviour, namely reduced sociability, repetition, obsessiveness and an inability to change its behaviour in response to other macaques. When they sequenced the macaque's DNA, moreover, they spotted two rare genetic variants linked to autism and other psychiatric conditions including depression, schizophrenia and bipolar disorder. The genes, known as *ABCA13* and *HTR2C* (their full names are a biochemical mouthful), reduced the number of neurons in the monkey's medial frontal cortex, a brain region important for social behaviour. In particular, the medial frontal cortex is known to house so-called mirror neurons – cells that let us mirror the behaviour of others – and so-called partner neurons – cells that let us respond appropriately to others. Yoshida's findings don't prove that monkeys can develop autism, but they nonetheless show that autism might not be an inherently human phenomenon.

The growing conclusion is that autism is with us for a reason. Our autistic ancestors probably played a fundamental role in shaping early human societies due to their unique strengths and special abilities. For example, many people with autism possess a deep understanding of natural systems. Isaac Newton, who experts agree showed signs of autism, was constantly preoccupied with understanding the systems that explained the complex motions of the planets. This obsessive focus on detail – called systematising – is how many autistic people are able to calculate what day of the week a given date falls on with astonishing accuracy: anything that obeys rules and patterns is child's play. Integrating such a skill into human communities would have enhanced our understanding of calendric systems and maps – critical innovations during times of environmental strife such as the wintry conditions of the Ice Age.

Because autism predominantly affects males, some scholars believe that systematising is proof of a popular theory of autism called the 'extreme male brain' hypothesis. It posits that autistics possess a turbo-charged 'masculinised' brain, and that higher than normal levels of testosterone cause autistic people to lack empathy. The idea was mainstream science until new research showed it to be bunkum. In 2019, in the largest study of its kind, a group of Canadian and American scientists found that the evidence for the male brain hypothesis had relied on samples far too small to prove cause and effect.[3]

What people with autism do often have, besides a remarkable memory, are enhanced vision, taste and smell. The incorporation of these skills into early human groups would have led to the creation of 'specialists': members who would have been indispensable for tracking, hunting and the creation of new farming systems – all of which would have been essential to group survival. Indeed, there is evidence for specialists in our own time. In 2005, Piers Vitebsky, an anthropologist who once lived among an indigenous community in the Russian Arctic, observed an elderly reindeer herdsman from Siberia who had memorised the parentage, medical history and character of every one of his 2,600 animals. Although he preferred the company of his reindeer to humans, he was highly respected in the community and had a wife, son and grandchildren.[4] His gift is just one example of the sophisticated, autism-like abilities of modern tribal peoples, who typically have at least one member with extraordinary skills of navigation, husbandry and predicting changes in the weather.

Autism is often accompanied by extraordinary visual skills. Some report that they can read the small print on products from across a room, or spot tiny variations in the fabric of a carpet. Temple Grandin, the famous autistic professor of Animal Science at Colorado State University, says her sight is so good that she often forgets to turn on the headlights when driving at night. The autistic brain processes fine detail much faster than the non-autistic, explaining why many autistics score twice as high as non-autistics on tests of visual acuity. From an evolutionary perspective, the benefits of enhanced vision are substantial: it helps us find food, avoid predators, seek out shelter and spot mates.

Art too may have benefitted from our autistic forebears. Cave art, such as the paintings found in the Chauvet Cave in southern France, demonstrate exceptional realism and a mystifying love of detail. Early humans relied on depicting and memorising their surroundings, and the autistic brain's ability to focus on parts rather than wholes would have made them the go-to choice of artist. Today, countless artists have revealed how autism has shaped their art. 'I became a geological and archaeological book illustrator using my skills in 3D and love of detail,' says John Adams, a digital media professional.[5] 'I have always been attracted to shapes and patterns of different forms and sizes and now create unique fused art glassware,' says Angus Corbett, an artist working with glass. 'Being an artist connects me to people. It gives me a language with which to share my unusual vantage point,' says Sonia Boue, an artist who has made films about art for Tate Britain, adding, 'My work is not about being autistic, but if you're touched by it and later discover I'm autistic, I may change your point of view.'

As for taste, it is well known that many autistic people are picky about what they eat and avoid trying new foods. Some will only eat mushy foods, for instance, and reject any food that doesn't fit a strict regime of only a handful of dishes. Since there are no differences in the taste buds of people with autism and the rest of the population, the trait must have its origins in the brain. It may be that the heightened sensory abilities of autistics make them oversensitive to certain tastes, like veritable supertasters. As for smell, studies have found that people with autism experience smell more intensely than is usual. Many will refuse to walk on grass because the smell is too overpowering. Others will refuse to change clothes each day because the smell of freshly washed laundry is intolerable. Why this is the case remains unclear; it may be due to a genetic variation that put their senses on a different evolutionary path. Whatever the reason, smell and the other senses are clearly different in people with autism. Some scholars think that heightened senses flood the autistic brain with information, interfering with their social brain and thereby prioritising perception over social ties.[6]

Attention to detail is a common trait in autism, beautifully captured in Mark Haddon's *The Curious Incident of the Dog in the Night-Time* when Christopher, the fifteen-year-old autistic narrator, describes an encounter with the police:

> Then the police arrived. I like the police. They have uniforms and numbers and you know what they are meant to be doing. There was a policewoman and a policeman. The policewoman had a little hole in her tights on her left ankle and a red scratch in the middle of the hole. The policeman had a big orange leaf stuck to the bottom of his shoe which was poking out from one side.[7]

This fondness for detail was long treated as yet another weakness by early researchers, who labelled it 'weak central coherence', a fancy term for not being able to see the forest for the trees, for missing the big picture.

But as researchers soon learned, focusing on the trees is no bad thing. In 1978 a famous study led by the psychologist Tim Langdell showed that autistics are much better at recognising faces than non-autistics, even if the image of a face is turned upside down or only the lower part of a face is shown.[8] They are better at seeing 'pure pattern' rather than 'social pattern', says Langdell.

This is a very useful skill for the Intelligence Community. Today GCHQ employs more than 300 staff with autism. '[They] have a different way of approaching problems,' said a member of GCHQ's recruitment team. 'They have a much more analytical, investigatory mindset.'[9]

Seeing pure pattern has another advantage. It stokes creativity, something autistics, long stereotyped as cold and inexpressive, were thought to lack. Ironically, this is because the tests researchers used to measure creativity were so uncreative themselves that they missed it entirely. They would ask a subject to name as many uses as they could for a paperclip or some other common object. Or ask them to use a circle to create as many drawings as possible in five minutes. Invariably, those with autistic traits would come up with fewer suggestions and not many drawings. But in 2015 scientists realised that while autistic people have fewer ideas, they are more likely to exhibit divergent thinking – thinking outside the box.[10] Common uses for a paperclip include a bookmark or a tool to clean your nails, but an autistic's uses for a paperclip include innovations such as a weight to balance a paper aeroplane or a light-duty spring. And where those without autism may use a circle to draw a smiley face or a pie, an autistic may draw a periscope view or a merry-go-round. The link between autism and creativity is staggering, and almost certainly reshaped human evolution.

Autistic creativity may have even reshaped our understanding of movement and light. In 2008, mathematicians spotted something oddly familiar in the Asperger artist Vincent van Gogh's *The Starry Night* (1889). The swirling patterns of luminescence in the sky seemed to be speaking their language. On closer inspection, and to their amazement, the painting displayed the mathematical formula of fluid turbulence.[11] Hidden within the dazzling eddies of stars is an equation strikingly similar to that discovered by the Russian mathematician Andrey Kolmogorov in 1941. No other artist's work demonstrates this. The physics of fluid turbulence is remarkably complex: it's thought to involve an interplay of tiny quantum whirlpools (small punctures in space-time) surrounded by cascades of circulating fluid. With no instruction, van Gogh's troubled mind had unlocked one of nature's deepest secrets.

Van Gogh's paintings contain many riddles that scholars still can't answer. But one thing is clear: the artist was not merely interested in painting, he was obsessed. It was an obsession that pushed him to the verge of insanity and probably foreshadowed his suicide in 1890 – a dark, inner pain that tormented him into manic depression and fits of creative brilliance: a curse and a gift. From an evolutionary perspective, autism and obsession share genetic roots. People with autism are twice as likely to be diagnosed with obsessive-compulsive disorder (OCD), and people with OCD are four times as likely to have autism. The overlap relies on a gene called *KDM4C*, an enzyme that regulates DNA. In the brain, circuits in the striatum (responsible for movement and reward) have been linked to people with autism and OCD, and a region within the striatum known as the caudate nucleus is unusually large in both groups. These brain regions also control habit formation, which helps to explain the repetitive behaviours seen in many autistics.

What's important to keep in mind is that obsessions can be a good thing. Though we often associate them with overpowering and unwanted thoughts, they can be an advantageous trait and have extraordinary results when channelled into skills such as tool-making, systematising thought, painting, or computer-coding. Today, we are dependent on the fruits of such obsessions. When the autistics working in Silicon Valley obsess over the latest technological innovations, amazing things happen that benefit us all. 'The movers and shakers have always been obsessive nuts,' wrote the American author Theodore Sturgeon. Obsessions can be empowering, life-changing and world-altering.

On a summer day in August 2021, Sally Hopkins looked up from her coffee to see me, once more, arriving for another chat about her extraordinary mind. My meeting with Sally, at a hip coffee shop in Shoreditch, was not to retread the familiar ground of how she experiences autism – I already knew a lot about her particular traits – but to learn what she thought about how autism aided human evolution. Since most of the research in this field was relatively new, I wondered how – and if – it changed the way she thought about her condition.

'It's amazing,' she said, her eyes beaming. 'It's as if there's been a de facto brain segregation in our society, a kind of neural apartheid, and it's finally crumbling.' Elaborating, Sally described how her unique skills and talents made sense in an evolutionary context. She counts things obsessively because our ancestors needed to systematise their thoughts. She draws dazzling wildlife pictures because our ancestors depicted animals in ancient cave art. And she possesses an all-consuming attention to detail because our ancestors needed to create complex tools to survive. Her mind is not a problem to be solved; it is the legacy of millions of years of brain evolution.

'I've been remembering things from my childhood in a completely new light – times when I was picked on, times when I knew my behaviour was okay even though I was made to feel it was wrong, times when my parents needlessly worried about me.' For a long time, Sally's parents didn't accept her autism. She still has difficulty discussing it with them.

When I asked Sally if she thought this new understanding would help dispel the myths about autism – that it used to be rare but is now common, that we're over-diagnosing eccentric children with an interesting condition, that we should be aiming to make people with autism indistinguishable from their peers – she was optimistic.

> I'm a big George Orwell fan, and one of my favourite quotes of his is, 'In a time of deceit telling the truth is a revolutionary act.' I think the autistic community has been living in a time of deceit for

centuries. It feels like it's been one myth after another, one discrimination after another. But now that we know that autism was a part of the story of how the human brain evolved, it feels like, by telling people this, it's our revolutionary act, our revolutionary truth.

Of Sally's many qualities, her intelligence struck me most. It's often said that nearly one in three autistic people has an intellectual disability, defined as having an IQ of below 70. But in reality, autism is a condition of high intelligence.

In fact, autism is now thought to have evolved several enhanced, but imbalanced, components of intelligence. One is the enlarged brain seen in many children with autism. Another is the well-documented synaptic peculiarities, which on one hand underlie difficulties in communication but on the other boost learning and memory. And the emerging picture is that, for all their difficulties with social interaction, autistics possess a stronger aptitude for focusing on tasks, a more deliberative decision-making style and a marvellous penchant for jobs involving science and technology.

Autistic children have even been found to get cleverer over time. Marjorie Solomon, a clinical psychologist at the University of California, Davis, and her colleagues have found that half of children with autism significantly improve their intellectual abilities between the ages of two and eight.[12] Even autistic children considered intellectually disabled can achieve average intelligence in the same period. 'This tells us that you can't be too hasty in diagnosing intellectual disability, because you just don't know what's going to happen,' says Solomon.[13] Her team analysed data from a long-term study of children diagnosed with autism, known as the Autism Phenome Project. In it, families are invited to have their child's brain scanned at age two or three, and then again two years later to assess their development. Behavioural and cognitive tests are also performed at this time, and then again when the child is between six and eight. From this, four groups of children emerge: the 'High Challenges', children whose intelligence drops over time; the 'Stable Lows', children whose intelligence remains the same over time; the 'Lesser Challenges', children whose intelligence increases modestly over time; and the 'Changers', children whose intelligence balloons over time.

Solomon's Changers represent the largest proportion of children in the study, a staggering 35 per cent. The Changers showed improved intelligence, better communication skills and a decline in disruptive behaviours. Solomon is now trying to understand what it is about the Changers that distinguishes them. 'We really think it's important to understand the differences, both biological and in terms of treatment experience and other characteristics,' she said. One unanswered question is why the Changers improved their verbal ability over time, yet only modestly improved in terms of 'autism severity': the behaviours associated with their condition. Curiously, it was the Lesser Challenges group that showed the greatest improvement in autism severity, despite not seeing the same intelligence gains as the Changers. Whatever the reason, Solomon concludes that her findings offer 'a hopeful message for many'.

The indisputable fact is that autism often brings new ways of seeing the world, and without autism it seems unlikely that human communities would be where they are today. This doesn't mean we should view ancient savants in a utilitarian fashion. 'Autism wasn't integrated solely because people with autism had some use,' Spikins points out, 'but because being human is about integrating people who are vulnerable without asking questions. And then a side-effect of that is that you do end up with all sorts of trades and talents that are useful.'

A pressing question for many is why we are only hearing about autism's role in evolution now. Why, given all our efforts to teach the public the truth about human origins and the wonders of evolution, has such a pivotal subplot remained conspicuously absent? The main culprit appears to be our evolutionary narratives, which tend to focus on the role of strong, independent men. Here again, the famous 'monkey to man' illustration that presents our history as a gradual transition from lowly ape to perfect man tells us more about the way we view ourselves than about human evolution. The final drawing is of a man, standing alone, ready to face whatever comes his way. It is an image sanctifying independence, the exact opposite of what made our species successful in the past and continues to do so today: interdependence.

'We still cling to this idea of complete independence,' said Spikins.

> Of the perfect, independent person who we can look back on and think, yes, that's us. When people ask me about human origins, what they often mean is *my* origins, the origins of *me* individually, what did *I* look like in the past? But when we look at the skeletal record there's virtually nobody who doesn't have some sort of injury or trauma or something that made them different from the rest of the group. Life was tough. We depended on other people and we weren't perfect. We were all sometimes vulnerable. And that just isn't in our narratives.

Neurodiversity

I dislike the word disability. It suggests there's something wrong with the individual rather than the society that excludes the individual. Most of us are only 'able' insofar as we have the necessary biological hardware for life as defined by the majority – adjustable limbs to navigate changing habitats, five basic senses to understand what's happening around us, and brains able to adopt the social and cultural conventions of the day. None of this would feel remotely special if an alien race with different needs colonised our planet.

Picture it: a hegemony of extra-terrestrial beings who use telekinesis to move objects and telepathy to exchange thoughts. If we were to coexist and thrive among such beings, we would have to rely on their willingness to modify their society. The alternative would be for them to decide against this, opting instead to label all of humanity 'disabled'. Perhaps our alien overlords would establish research centres to investigate what was wrong with us. Perhaps they would offer us medication to be more like them. The point, wonderfully made by an autistic, pseudo-named 'Muskie', is that this is precisely what non-autistics, otherwise known as neurotypicals, have been doing to autistics for decades. In a playful tit for tat, Muskie inverts the narrative with his website 'The Institute for the Study of the Neurotypical', which states:

> Neurotypical syndrome is a neurobiological disorder characterised by preoccupation with social norms, delusions of superiority, and obsession with conformity . . . NT is believed to be genetic in origin. Autopsies have shown the brain of the neurotypical is typically smaller than that of an autistic and may have overdeveloped areas related to social behaviour.[14]

You don't have to be a neuroscientist to understand that autism is merely a different kind of mind. In 1993 the autism-rights activist Jim Sinclair wrote the famous essay 'Don't Mourn for Us', calling for an end to the fear, pity and public stigma associated with autism. Parents in particular were reprimanded for their role in creating the narrative. In words controversial to this day, he declared,

> You didn't lose a child to autism. You lost a child because the child you waited for never came into existence. That isn't the fault of the autistic child who does exist, and it shouldn't be our burden. We need and deserve families who can see us and value us for ourselves, not families whose vision of us is obscured by the ghosts of children who never lived. Grieve if you must, for your own lost dreams. But don't mourn for *us*. We are alive. We are real. And we're here waiting for you.[15]

Sinclair's mission was to allow people with autism to become the narrators of their own experiences, free to validate their feelings and ready to celebrate their differences as all other differences are celebrated. Following Sinclair, the Australian sociologist Judy

Singer coined the term 'neurodiversity' in 1998, which was quickly picked up and championed by the autistic community. In a nutshell, it means the following: just as there are infinite variations of body type, ethnicity and culture, so too are there infinite variations in neurocognitive function within our species. Autism is therefore just another form of human diversity. There are no 'misfits', 'oddballs', 'eccentrics', 'nerds', 'loners' or 'weirdos' – there is only humanity, with all the wondrous biodiversity and neurodiversity it entails. In her pioneering sociological thesis, Singer declared,

> I wanted to see a neurodiversity revolution as potent as the feminist revolution had been. I wanted to see if, given a more understanding, inclusive and supportive environment . . . a new type of human, capable of rising to a new level of human creativity, would evolve.[16]

Reaching this understanding has not been easy. Even today, the American Psychiatric Association's *Diagnosis and Statistical Manual of Mental Disorders*, the *DSM-IV*, defines autism as a condition manifesting in an 'abnormal development in social interaction and communication, and a markedly restricted repertoire of activity and interests.' But I am reluctant to accept the words of a psychiatric organisation that as recently as 1968 defined homosexuality as 'a mental disorder'.

The sad truth is that for many autistics, the medical establishment – with its need to atomise, objectify and institutionalise – has been an engine of oppression, not emancipation. As Temple Grandin observes:

> Parents come up to me all the time and say things like, 'First my kid was diagnosed with high-end autism. *Then* he was diagnosed with ADHD. *Then* he was diagnosed with Asperger's. What is he?' I understand their frustration. They're at the mercy of a medical system that's full of label-locked thinkers. But the parents are part

> of that system too. They'll ask me, 'What's the single most important thing to do for an autistic kid?' Or 'What do I do about a kid who misbehaves?' What does that even *mean*?[17]

Grandin, who has had her own brain imaged and studied, goes on to explain that the differences are what make us individuals. For example, she has more connections in her corpus callosum (the bundle of nerves bridging the left and right hemispheres) but others could just as well have more, fewer, or the same number. Her brain's language circuits branch more than a 'normal' person's, but this too exists on a continuum of neurological variation. The same is true for her unique genome. 'I have often thought that eventually we're going to be asking ourselves at what point this or that autism-related genetic variation is just a normal variation,' she writes. 'Everything in the brain, everything in genetics – they're all on one big continuum.'

An unfortunate product of autism research, in my view, is the Autism Spectrum Quotient (AQ).[18] Like IQ, AQ seeks to label people based on some pretty sketchy science. The test consists of a fifty-item questionnaire meant to assess things like social skill, attention to detail, communication and imagination, scored on a 4-point scale (1 = definitely agree to 4 = definitely disagree). The average score is 16.4, with scores above 30 veering into autism territory. Examples from the questionnaire include: 'When I'm reading a story, I can easily imagine what the characters might look like'; 'I would rather go to the library than a party'; 'When I talk, it isn't always easy for others to get a word in edgewise'; 'I enjoy doing things spontaneously'; 'New situations make me anxious'.

Frankly, they're absurd. Anyone could be classified as autistic or non-autistic with questions this vague. We can all get pretty anxious in new situations, and I, for one, enjoy going to the library more than a party. The test's goal, of course, is to create arbitrary lines to label-lock people into arbitrary groups. Perhaps you scored 'better' in the social domain, but 'poorly' in the imagination domain. Perhaps, based on your score, you do not have the 'deficit' of autism after all.

Temple Grandin points out that AQ can shame many autistics into silence: 'A generation ago, a lot of these people would have been seen simply as gifted. Now that there's a diagnosis, however, they'll do anything to avoid being ghettoized.'

Cracks in AQ started to appear in 2017 when Swedish researchers showed that the majority of the questions (thirty-eight out of fifty) had no explanatory power. They couldn't even suggest who did and who didn't have autism.[19] Worse, we now know that even if a person scores high on the test (over 32), the odds of that person having autism are about one in ten. When scientists tested AQ using what we call the positive predictive value, which is the probability that someone with a positive test result truly has the condition, the data churned out only 8 to 12 per cent, hardly a ringing endorsement. Ironically, the failure of AQ to spot and measure autism revealed more than the test itself ever could.

The public's understanding of autism is certainly improving. Progressive writings such as Steve Silberman's *Neurotribes* and Andrew Solomon's *Far from the Tree* have changed the zeitgeist, shifting society away from biological determinism and towards a more liberating social constructivism. Consequently, the neurodiversity movement has spread far and wide, creating information networks that allow autistics to share their experience and enter mainstream society. Autistics are finally starting to see themselves for who they are: a diverse community of extraordinary individuals with unique brains, long oppressed by a dominant culture of unenlightened, unchallenged neurotypicals. But don't take my word for it. Hear what they themselves have to say:

> ADAM: I'm not weird or socially awkward, I'm just different. I like to be logical and I don't understand why people lie. My autism comes with challenges, but I definitely don't see it as a burden. It's a part of me and I wouldn't change it . . . I would change how

others view me. I see it in their eyes, that 'something's wrong' look. And I want to say to them, 'My brain just works differently, please don't judge me.' But I feel too sad and frustrated to say anything . . . When will people accept us for who we are?

SUSAN: I feel that people should stop trying to fit autistics into neurotypical moulds. As a lower functioning autistic who can't tolerate the strain of hard effort and rigorous learning, I feel that people should stop pushing me to 'train myself out of my shortcomings' . . . if autistics and other disabled persons want to push themselves to the limit and put themselves through tough teaching programmes, fine with them. But those who can't shouldn't be put through torture . . . I am fed up with being thought of as lazy . . . I wish they'd realise that some people may not have the wiring necessary to do certain things.

GWENN: Neurotypicals need to get over themselves. Why should I bend to their ways? How about they try to be more like me? I learnt to read before all my classmates. I see pictures with every new thought. But all I hear, all the time, is how 'difficult' I can be. How science needs to find a cure for me. To be honest, I think it's appalling. I know there are autistics out there who want and sometimes need medical help, and I wouldn't dream of speaking for them, but we need to stop lumping all autistics together. Treating me would feel like assault.

LEE: When I discovered that others like me were speaking out about their autism, I couldn't believe it. I'd been living completely under the radar. I wouldn't even *try* to talk to people about it, because as soon as the 'A' word comes up the conversation moves to a parallel universe, a nightmarish universe, where every word, every syllable leaving my mouth is dissected by the other person's thoughts. I know they're trying to be supportive, but they can't help it: society has programmed them to see us a certain way, to feel sorry for us . . . We will one day reach a true understanding. I'm optimistic now. The neurodiversity movement has changed everything.

RAMILA: We need to organise ourselves for our voices to be heard. Neurodiversity is a great start but its message will only be effective if it changes the public's perception of autism. I'm always online now, searching for others who will come out of the closet, so to speak . . . I don't think anyone with autism should feel scared or stigmatised by society. I think society should feel ashamed of the way it's made us feel. I think society is the thing that needs to change. We're here now. And more of us are being found every day.[20]

Could autism be the next phase of human evolution? With autism on the rise, the question isn't nearly as absurd as it sounds. Today's 1 in 59 rate represents a 15 per cent increase from the 1 in 68 rate of 2010, an 86 per cent increase from the 1 in 110 rate of 2006, and a 154 per cent increase from the 1 in 150 rate of 2000. Some of this rise will no doubt be due to increased awareness and better ways of identifying autism, but not all of it will. Which leaves the door tantalisingly open.

This is what I love about evolution. Autism *could* be the next phase of human evolution. It's entirely possible, probable even, if just two conditions are met. The first is that autism must be heritable, that is, transmissible from parent to offspring via genes. Which it is. In fact, recent estimates place the heritability of autism at 83 per cent, up from a previous estimate of 50 per cent. This means that the child of an autistic parent is highly likely to develop some form of autism themselves. The second condition is that people with autism must have differential reproductive success, which is a fancy way of saying they need to have more children than people without autism. This isn't so clear. Mild forms of autism can go unnoticed, and we're a long way from understanding how wide the spectrum really is. Yet there are signs that people with autistic traits are reproducing more than they once were. In 2014 researchers demonstrated that children considered 'geeks' consistently outperformed those deemed 'cool' in later life.[21] Cool kids were more likely to develop drug addiction and social insecurities, while geeks and social introverts, many of whom very probably had some form of autism, were more likely to thrive in their careers and go on to form meaningful relationships.

Evolution has no purpose, of course, so there's no reason the autistic brain would be singled out specifically. Despite the beguiling high-functioning traits of many autistic minds – the enhanced memory, learning and intelligence – it would be a fallacy to suggest that these brains are in competition with neurotypical brains for ascendency. Yet there is something about different minds and our reaction to them that feels purposeful. One of the most singular traits that has evolved in the human brain is empathy, the capacity to perceive and apprehend another's misfortune. In decades past our empathy for autistics was virtually non-existent. Autistics were labelled as mentally retarded or insane, locked up in asylums and underwent medical interventions that were as thoughtless as they were cruel. But empathy has experienced its own form of evolution. I want to say we're now more compassionate towards those with different minds, but even that feels condescending. More apt, I think, is to say we're finally *alive* to those with different minds. We recognise that such diversity is to be treasured, not problematised – embraced, not feared.

Reaching this understanding requires every faculty evolution has built for us: our ability to form groups and build social bonds; our ability to construct emotions right for the times; our ability to remember one another's plight and lay down collective memories; and our ability to marshal our astonishingly malleable intelligence and ever-expanding repertoire of language.

Rethinking autism is a grand project. So it's encouraging that new ways of conceptualising it are beginning to emerge. Among them is how we think about stimming, the repetitive fidgeting behaviour seen in those with autism (rocking, bouncing, flapping etc.). Many autistics say it calms them down when they are feeling anxious and allows them to express their excitement about something when words fail them. Unsurprisingly, stimming has long been discouraged and even suppressed by parents and teachers who view it as disruptive and potentially harmful. One girl with autism reports how she was forced to cross her fingers to stop snapping them, only to find that she then couldn't do her schoolwork. One boy with autism reports how he was forced to hunch over his desk to stop rocking back and forward, only later to develop back problems and a poor posture. With children like these in mind, activists in Canada have launched the International Day of the Stim, a day for all autistics to share their stimming experience with others on social media. 'It was really an example of autistics reclaiming something that, through negative and abusive therapies, had been taken away,' says Anne Borden, an activist who helped establish the day.[22]

Another positive sign has come from the tech giant Microsoft, which recently announced that neurodiversity would become central to its recruiting strategy. And where Microsoft goes, others usually follow: now multinationals including Ford, Ernst & Young, J. P. Morgan Chase and the Royal Bank of Canada have started neurodiversity employment programmes. If companies like these can persuade others of the need for neurodiverse people in the workplace, and if the wider public can pressurise governments to enforce neurodiversity quotas at state level (a recent poll found that 72 per cent of UK employers ignore neurodiversity in their policies[23]), the movement would spread rapidly.

No one knows how many autistics are out there, lost, hidden in the shadows of prejudice and ignorance; some estimate it may be up to 200,000 in the UK alone.[24] Only when we change our own minds will we recognise the wonders of the autistic mind. Autism is not and should never be something to solve. It is one of the most beautiful spin-offs of human brain evolution.

9

The iBrain

With all the great achievements of humanity, it is tempting to think that the human brain may be at the pinnacle of its evolution and development. From our vantage point, our ancestors seem like children scrambling up a hill, while we, the 'grown-ups', smugly gaze down from the mountaintop.

But in fact our brains are still evolving – perhaps faster than ever. Predicting where this evolution will take us next is a question for all scholars in all disciplines, for the very nature of brain evolution is also changing. New technologies such as brain–computer interfaces, artificial intelligence, reprogrammable stem cells, and genetic tools such as CRISPR – a new technology that allows scientists to make precision edits to DNA – may forever change what it means to be human.

First the bad news. It is only a matter of time before *Homo sapiens* go extinct. All species go extinct sooner or later. Over 99.9 per cent of all species that have ever lived are extinct, and there is no reason to think that our species will somehow be exempt from this sobering reality. Even if humans avoid a mass extinction event – a lethal pandemic or a massive asteroid – they will still branch into a new species of human, meaning the original *Homo sapiens* will have gone extinct. In short, evolution and extinction are not mutually exclusive: when speciation occurs, it means that selective pressures have driven the older species to extinction. Though physicists and others often dream about colonising Mars and other planets, personally I think this is the only planet we have. I might be wrong. I hope I am, given how destructive *Homo sapiens* can be.

The good news is that our descendants will live on. Just as earlier species of humans (including *Homo habilis*, *Homo heidelbergenis*, *Homo neanderthalensis* and *Homo floresiensis*) and other great apes (including chimps, gorillas, bonobos and orangutans) branched into different species, *Homo sapiens* will split into several new kinds of ape entirely. In his novel *The Time Machine*, H. G. Wells foretold that by the year AD 802,701 humanity would evolve into two separate species, the blissfully naive Eloi and the grisly cannibalistic Morlocks, both in zero-sum competition with each other. While some experts predict a fate similar to Wells's dystopia, with humanity splitting into a genetic upper class and underclass, I am not so pessimistic.

A new movement called transhumanism is taking root in our world, which aims to upgrade humans with gene enhancement and tech implants to such an extent that we become different beings entirely. Posthuman beings. When old age creeps up on us, we'll be able to turn back the clock using an app that reprogrammes our DNA. When we want to go hiking or deep-sea diving, we'll simply tweak the genes that make our muscles strong and our lungs hold breath for longer. And if these upgrades aren't good enough, technology will step in. Body augmentation devices such as contact lenses that take pictures or videos, language translators that allow us to speak to anyone in the world, and mind-controlled prosthetics that let us pay for our shopping and share our holiday experiences with a nod of the head will become commonplace.

These are not fantasies of science fiction. In 2017 scientists established the first academic journal dedicated to pursuing transhumanism, the *Journal of Posthuman Studies*. 'As the boundaries between human and "the other", technological, biological and environmental are eroded,' the journal proclaims, 'and perceptions of normalcy are challenged, they have generated a range of ethical, philosophical, cultural, and artistic questions that this journal seeks to address.' As our brains evolve into those of a new type of human, answers to such questions will no doubt raise disturbing questions of their own. Would immortal brains keep regressive thinking alive forever? (Now there's a terrifying thought for you.) Would post-human brains lead us into a tech-obsessed dystopia, where all creativity and diversity of thought is obliterated?

Of course, the fear of science and advanced technology will probably always be with us. Steven Pinker calls it 'progressophobia'. When physicists built the Large Hadron Collider (LHC), the world's largest and most powerful particle accelerator, many feared it would generate a black hole so large it would devour Earth and all life like a cosmic Cookie Monster. Only after years of public engagement have physicists allayed the public's concerns. Brain advancements are perhaps more unnerving because they strike at the core of what makes us human. Even I – a brain-obsessed technophile – think there's something quite chilling about a computer knowing what I'm thinking. Nevertheless, the greatest advances have already begun.

Theseus's Paradox

In ancient Greek mythology, Theseus, the Athenian king who slayed the Minotaur, was so revered that his ship was preserved for hundreds of years. When the ship's wooden planks started rotting away, they'd simply be replaced with new planks made from the same material. When the entire ship's timber had been replaced, however, philosophers asked the logical question: can it still be considered Theseus's ship? This paradox is now a major talking point in neuroscience labs, where it leads to heated debates about identity and the self.

Consider the following: it is the distant future, mortality is a thing of the past, and stem cell technology has reached its zenith. The molecular mechanisms of memory are completely understood, and when our brains age and our memories fade, both are simply replaced. Imagine you did this for hundreds of years. Would there come a point when your past self is no longer considered you? Would you become like the ship in Theseus's paradox?

Like it or not, our species will eventually have to confront this question. Jürgen Knoblich, director of the Institute of Molecular Biotechnology of the Austrian Academy of Sciences in Vienna, is at the forefront in experimenting with this concept. In 2013 Knoblich began growing large parts of the human brain in a laboratory dish.[1] These 'organoids' are still in their early development – and are not, Knoblich insists, 'humanoids in a jar'. But the notion of using organoids to generate everlasting spare parts for the brain is not far off; indeed, monkeys suffering from neurological disease have had parts of their brain replaced with laboratory-grown human brain cells.[2] 'The idea of growing a human brain in a jar has always fascinated people,' Knoblich said in an interview following his breakthrough:

> There are tons of science fiction movies that deal with that. The real use of our system, though, is that we can grow brain tissue essentially from any human, whether healthy or diseased. And that raises the hope that some of the major neurological disorders eventually can be modelled in this system.[3]

I've long been fascinated with stem cells. In my *In Pursuit of Memory: The Fight Against Alzheimer's*, I devoted a chapter to exploring their potential to treat and perhaps even reverse the symptoms of Alzheimer's, a condition I became intimately familiar with when I witnessed my grandfather succumb to the disease. In 2012, when he was nearing the end, his treatment options were still as poor as they were in the 1980s, so the idea of growing brain tissue in a dish to create an endless supply of neurons was tantalisingly promising. For a time they were my biggest hope, my deepest obsession.

But I now realise I wasn't thinking big enough. As we saw in our discussion on memory (chapter 4), the human mind hasn't evolved to construct libraries of past selves tucked away in some basement of the brain. We are our memories as much as a painting is its brushstrokes. In the future, then, the miracle of neural stem cells won't be their ability to preserve what we were, but to maintain what we are.

And this will rely on our brain's ability to self-repair. In 2019, scientists at the University of Plymouth made a breakthrough with this concept. They found that one of the skills of neural stem cells is their ability to 'wake up' neighbouring brain cells by releasing a package of molecules called STRIPAK (Striatin-interacting phosphatase and kinase).[4] STRIPAK has evolved in countless organisms from fungi to mammals and has many jobs: it helps move things around inside neurons, it helps other cells in the body divide, it even helps our hearts function. If scientists can pin down what it is about STRIPAK that rouses our brains stem cells into action, there may be no need for Knoblich's organoids after all. Unleashing the brain's healing powers not only protects us from disease; it makes us smarter as well. Stem cells in the hippocampus have a remarkable effect on learning and memory, both of which are linked to intelligence. And while the adage 'use it or lose it' is true, because your brain really is like a muscle, it's also true to say that if you don't lose it, you can always use it. In this way, rejuvenating rather than replacing our brains may be the true future of brain repair.

Organoids and self-repairing brains are not the only brains of the future, of course. Could artificial intelligence (AI) brains evolve in a way similar to our own? Many neuroscientists think they could. That's because the algorithms that underpin machine learning are slowly being replaced by what we call genetic, or evolutionary algorithms. Inspired by Darwin's theory of natural selection, evolutionary algorithms work by mimicking the features of a genome, introducing artificial mutation and recombination events into a computer's software until the 'fittest' programme is selected.

Genetic algorithms are already being used in AIs such as Google's DeepMind and AlphaGo, the latter of which recently beat the Go world number one, Ke Jie, from China. After his defeat, Jie called the programme the new 'Go god', vowing never again to subject himself to the 'horrible experience' of challenging the AI system.[5] Genetic algorithms are the first step towards an AI brain matching that of a human. In October 2017 the neuroscientist Stanislas Dehaene of the Collège de France in Paris proposed two additional steps, which he calls consciousness 1 (C1) and consciousness 2 (C2).[6] C1 is our ability to maintain a diverse range of thoughts simultaneously, making faculties such as forward planning possible. C2 is our capacity for introspection, allowing us to reflect on mistakes and improve over time. If these steps are realised, Dehaene writes, AI 'would behave as if it were conscious . . . it would know that it is seeing something, would express confidence in it, would report it to others . . . and may even experience the same perceptual illusions as humans.'

Conscious AI terrifies most people. The great worry is that it would escape our control and destroy the human race. Legendary scientists including Alan Turing, Stephen Hawking, Bill Gates and Tim Berners-Lee have all voiced grave concerns about its potential. And yet their fears remain abstract for good reason: we're still not sure what we mean by 'conscious' AI. As we saw in chapter 7, all forms of consciousness are really just illusions to position organisms in their ecological niche. But what's a machine's ecological niche? An air-conditioned server room in the Pentagon? Your office desk? Your phone pocket? The fact is, computers do not operate in the territory of what it means to be human. A machine has to want something beyond self-replicating programmes and self-preservation to be in the same league as us. Human brains create meaning, not computation: they reflect the lives they have lived and the world they have experienced. They construct rather than represent information. So to reduce their wondrous complexity to a computer code is, in my view, a mistake. Perhaps the more realistic future lies in a combination of mind and machine.

Homo Cyber Sapiens

'Any sufficiently advanced technology,' said the futurist and writer Arthur C. Clarke, 'is indistinguishable from magic.' This is especially true of a brain–computer interface (BCI) called the BrainGate system: a series of sensors implanted in the motor cortex, which allow a person to control muscles in their arm using the power of thought. John Donoghue, the founder of BrainGate, believes that such technology can help people who have had a stroke, lost limbs or been paralysed by diseases such as amyotrophic lateral sclerosis (ALS) and locked-in syndrome (LIS). LIS received world-wide attention when the French journalist Jean-Dominique Bauby dictated his experience of the disease by blinking one eye to write his famous book *The Diving Bell and the Butterfly*, in which he likens his paralysed body to a diving bell, with his mind, a butterfly, trapped inside. Ever hopeful of future therapies, he longed for the day science would free his mind: 'Does the cosmos contain keys for opening my diving bell? A subway line with no terminus? A currency strong enough to buy my freedom back? We must keep looking.'[7] Donoghue is convinced his BrainGate system would have helped Bauby: 'I would have every expectation that if we had put BrainGate in his brain, it would have immediately started giving us signals.'[8]

Many people are now using BCIs. In 2017, researchers at the University of California, Irvine, used BCI technology in a paraplegic man to restore his walking after a spinal cord injury.[9] Other researchers are using BCIs to stimulate the visual cortex to treat the blind,[10] restore lost connections in stroke victims,[11] and scan the brain for signs of depression. Treating the mind with machines dubbed *electroceuticals* has helped hundreds of thousands of people with hearing loss – using cochlear implants – and thousands of people with Parkinson's disease and epilepsy – using deep brain stimulation (DBS). DBS is particularly impressive because it uses tiny electrodes implanted in the brain which send electrical impulses to regulate brain activity, all powered by batteries sewn into the patient's chest. The concept itself isn't new. The ancient Egyptians used electric catfish to treat arthritis, and the Romans used electric rays to treat headaches. But the technology is thrilling because it represents the first step towards creating *bionic* brains.

It is no exaggeration to say that such technology is allowing humans to shape their own evolution. The first to do exactly that is a man named Neil Harbisson, the world's first officially recognised cyborg (short for cybernetic organism): a being that is part human, part machine. Born in Belfast and brought up in Spain, the thirty-seven-year-old has a rare genetic condition called achromatopsia, or complete colour-blindness: he sees the world in shades of grey. If you were to bump into Harbisson, one of the first things you would notice is the long black antenna protruding from his head, which translates wavelengths of light into musical notes. This gives Harbisson extrasensory perception, and he can 'feel' colours inside his head.

As an artist, Harbisson uses his extra sense organ to create dazzling technicolor performances and exhibits. One of his specialities is sound portraits, an artform in which he stands in front of a subject and uses his antenna to record the different 'notes' emitted from their face. He then downloads this onto a sound file, producing a facial portrait you can hear. His work has become so popular that A-list celebrities including Robert De Nero, Leonardo DiCaprio and Woody Allen have had sound portraits made. Describing himself as the first 'trans-species' person, Harbisson feels intimately connected with his neural prosthetic, telling *National Geographic*, 'There is no difference between the software and my brain, or my antenna and any other body part. Being united to cybernetics makes me feel that I *am* technology.'[12]

Such progress will take time to be accepted. In 2004, the UK Passport Office rejected Harbisson's passport photograph, saying that it should contain 'no other people or objects. No hats, no infant dummies, no tinted glasses.' After weeks of Harbisson explaining that his antenna was simply 'an extension of his brain', the baffled officials issued his passport. Seven years later, during a demonstration in Barcelona, Harbisson was attacked by three police officers who thought he was filming them. When he explained that he was a cyborg and was simply listening to colours, they laughed and tried to pull his antenna off his head.

The concept of the cyborg dates back to at least 1839, when the writer Edgar Allen Poe published his short story 'The Man That Was Used Up'. The plot follows an unnamed narrator who meets the formidable General John A. B. C. Smith. Six feet tall and powerfully built, Smith is also a skilled raconteur and possesses a host of other enviable characteristics. But our shrewd narrator is convinced the general has a secret and, when he visits his home, he learns what it is. Piled on the floor is an assortment of strange objects which speak in the voice of the general. A servant then enters the room and begins piecing the objects together – an eye socket here, a leg there – slowly revealing the general to be no more than an assembly of artificial prostheses.

Poe's tale certainly captured the public's imagination, but it was the American scientists Manfred Clines and Nathan Kline who coined the term cyborg in 1960. Writing in the September issue of the journal *Astronautics*, they declared (in wildly ornate language), 'For the exogenously extended organizational complex functioning as an integrated homeostatic system unconsciously, we propose the term "Cyborg".'[13] Their dream was to create human–machine hybrids to enable humans to explore outer space. Only then would we transcend nature and finally control our own evolution. 'Space travel challenges mankind not only technologically,' they wrote, 'but spiritually, in that it invites man to take an active part in his own biological evolution.'

Another great advance is the Hybrid Assistive Limb (HAL). This is a wearable robotic exoskeleton suit, designed in 2012 by Japan's Tsukuba University and the robotics company Cyberdyne, to help those who cannot walk due to spinal injury. Using sensors attached to the skin, the HAL suit boosts the electrical signals sent down the spinal cord from the brain, bypassing the injury and allowing the person to activate their muscles. The HAL suit then feeds back to the brain, teaching it what signals are necessary to walk: the first step towards walking without being assisted by the HAL. Scientists are now exploring the potential of HAL to treat stroke, paralysis and missing limbs.

The aspirations of BCIs do not end there: Silicon Valley companies such as Facebook and Google are now investigating the possibility of thought-to-text typing, neural processors for enhanced concentration, decision-making and fitness, and even ways to record and download our dreams. Imagine it: all those whirling thoughts, feelings and sensations instantly captured and played back for analysis. Most psychologists have given up trying to interpret dreams, but with that much data I suspect we could readily interpret them ourselves. We're not there yet, but neural interfaces are currently able to detect how we respond emotionally in different situations. This so-called 'mood reading' has piqued the interest of those in the 'neuromarketing' world, ever watchful for technology that can detect how people respond to products and advertisements, as well as monitor their employees' moods.

One of life's greatest bugbears is our struggle to put thoughts into words. More often than not, we have to see our thoughts written down by a gifted journalist or hear them articulated by a skilled orator. We know what we think; we just need someone else to say it. As the famous technologist Mary Lou Jepson explains, 'The [brain's] input's pretty good, but the output is constrained by our tongues and jaws moving and us typing . . . If we could communicate at the speed of thought, we could augment our creativity and intelligence.'[14] In 2019, researchers at the University of California, San Francisco took us a step closer to this reality, creating a device that converts brain activity into speech by decoding the signals the brain sends to the tongue, lips, jaw and throat. The device can even decode speech when a person silently mimes sentences.

Researchers at Kyoto University, moreover, have successfully decoded the brain activity of someone looking at an owl, and when they converted this signal to a computer screen, the image looked very similar (if a little blurry) to the owl the person saw.[15] In Charlie Brooker's science fiction show *Black Mirror*, in the episode titled 'The Entire History of You', people possess 'grains': neural implants that record everything they do, see or hear. They can then rewind to any memory they choose, and even project the image onto a screen. The episode takes a dark twist when the main character, Liam, uses the technology to discover that his wife had an affair. Like other episodes in the series, it questions how technology will change the rules of society. It is no exaggeration to say that the Japanese researchers' owl experiment could be the first step towards that future.

At the University of Washington, Seattle, the neuroscientist Andrea Stocco has taken things even further, showing that two brains can be directly linked across the Internet.[16] This mind-to-mind communication allows each participant to know what the other is thinking. 'The Internet was a way to connect computers, and now it can be a way to connect brains,' says Stocco. In Stocco's experiment, two people, each in a room a mile apart, wear an electrode-studded cap connected to an electroencephalography (EEG) machine to pick up signals from the brain. They then play a guessing game in which one person thinks of an object and the other person has to guess what it is. Remarkably, Stocco's subjects arrive at the correct answer 72 per cent of the time. 'We knew in theory it could work,' he said. 'Now we want to discover how well it can work.'

Like the 'mindmelds' of *Star Trek*, mind-to-mind communication could allow humans to become telepathic – transmitting thoughts between two brains without ever having to utter a word. Though the experiments performed thus far have involved transmitting very simple information such as object guessing and binary choices (left or right, one or zero), it's only a matter of time before more complex messages are sent telepathically. The days of emailing and texting would be over. Instead, humans would simply beam their thoughts to each other.

One of the biggest contributors to neural interface research is the military. The Defense Advanced Research Projects Agency (DARPA), America's nerve centre for new military technology, is now developing brain–computer interfaces that can sharpen soldiers' mental skills, help them see more effectively in the dark, and allow them to control swarms of drones at the speed of thought. Some experts say this technology is a cleaner, safer alternative to the prescription drugs used by soldiers to boost performance, such as modafinil and Ritalin. Others, though, are worried about the long-term effects on a soldier's psyche, and how such advances would fundamentally change the theatre of war: killing the enemy might suddenly feel as fictitious as shooting the bad guys in a video game. Then again, perhaps the precision of thought-operated weaponry would act as a better, less destructive deterrent than nuclear weapons.

While these advances may seem to be primarily technological, they actually represent the next step in our brain's evolution. Many neuroscientists think that our brain is already operating at maximum capacity, and that any improvement in intelligence and information-processing is constrained by the number of neurons we possess. (Among the most popular of brain myths is that we only use 10 per cent of our brain's capacity; in reality, the entire brain is constantly active, even during sleep.) So unless we find a way to create neurons externally (something we're not even sure is possible), this means the next step for our brains may not be the organic evolution we have witnessed thus far. Indeed, the futurologist Ian Pearson envisages a future in which humans have merged with machines effectively to become a new species, called *Homo cyberneticus*. As this species develops increasingly sophisticated brain–computer interfaces, they evolve into *Homo hybridus*, until the organic parts of our brain are completely replaced by machines, leading to the rise of *Homo machinus*: a brain made entirely out of silicon, allowing our minds to achieve digital immortality.

The question on everyone's mind, of course, is when? What timeline can we realistically expect for such radical brain advances? To answer that we must look to history. It took little more than a century to go from the horse-drawn carriage to NASA's Mars Exploration Rover, and, if we follow Moore's law – which states that the speed and capability of a computer doubles every two years – brain–computer interfaces may be at the stage computers were at in the 1970s. Which means that many of the advances discussed so far could become a reality in the twenty-first century. I suspect thought-to-text typing will come out just around the time I shuffle off this mortal coil. Damn.

Sceptics among you might point out that we surely need to understand the brain for such technologies to work, and we are certainly a long way from that understanding. The brain is endlessly, dizzyingly, stupefyingly complex; if I had to quantify it, I would say that we comprehend less than 1 per cent of its functionality. Yet a fuller understanding isn't always necessary. The entire computer industry is built on quantum mechanics, a field just as complex and bewildering as neuroscience. So too are lasers, telecommunications, atom clocks, GPS, and Magnetic Resonance Imaging (MRI). Whatever the scientific discipline, there will always be levels of understanding beyond our current grasp.

Some are unsettled by this evolutionary future and the ethical issues that neural interfaces raise. If brains can be reached on the Internet, they can also be hacked. In 2019, scientists at the UK's Royal Society scrutinised the risks of BCIs, questioning how governments and tech giants would – or should – control access to neural interfaces, and whether there should be limitations on their use. 'If neural interfaces can be voluntarily used to influence behaviour by individuals, then should these be prescribed by states?' they ask.

> For example, should technologies that seek to help people eat more healthily be used by governments to reduce their public health bill? Then, should that power be extended for use in wider contexts, for example as sanctions in criminal justice? Conversely, in terms of proscribing applications rather than prescribing them, should there be 'red lines' beyond which interfaces are banned? And who would decide where such lines would be drawn?[17]

No one, for now, has any answers.

Another concern for the Royal Society report authors is that if we start using machines to perform brain functions, 'is it us as humans doing it? Or is it the technology?' Human agency, our capacity to make choices and act on those choices in the world, is among the most important features of our existence. We enshrine it in law and state institutions. It is the backbone of human freedom. One could of course argue that neural interfaces enhance human freedom by making life easier; after all, spending hour after hour concentrating, solving problems, and struggling to express oneself is hardly liberating. (Just think what you could do with all that extra time.) And yet, there is something distinctly human about overcoming life's challenges. A fortitude that gives meaning to our lives and one that we might be unwise to forsake.

Whatever happens in the years ahead, the fact is that brain technology will change everything forever. As I write, all manner of neurotechnology is being developed, with sci-fi names such as neural lace (a mesh of tiny electrodes implanted in the skull to monitor brain function), neural dust (a wireless nerve sensor, powered by ultrasound, to monitor and control nerves) and neuropixels (probes that can record the activity of hundreds of neurons). As neuroscientists and others continue to think of ways to make machine-enhanced minds a part of our world, such advances appear, to many, to be a step too far. Nowhere does this debate seem more urgent than on the subject of human violence.

Rewiring for Peace

In the early afternoon of 24 February 1999, Donta Page, a career burglar, broke into the home of a twenty-four-year-old charity worker named Peyton Tuthill in Denver, Colorado. Terrified, Peyton ran upstairs – but Page caught her. After punching her several times in the face, he dragged her into a bedroom and raped her. He then slit her throat and stabbed her six times in the chest. According to his taped confession, Page murdered her because it was 'just the first thing that came to mind'.

This type of senseless violence is not unique to our species. Chimpanzees display psychopathic behaviour. Dolphins rape and murder each other for fun. Nature, wrote Tennyson, is 'red in tooth and claw'. No one knows what the evolutionary roots of crime are. In the 1870s Cesare Lombroso, an Italian criminologist and physician, became convinced that criminals are a devolved, undeveloped form of our species. He made curious biological observations such as this: 'A criminal's ears are often of a large size. The nose is frequently upturned or of a flattened character in thieves. In murderers it is often aquiline like the beak of a bird of prey.'[18] These observations were erroneous, of course, but the latest research suggests that certain brain attributes may incline some people to violence.

A new field known as neurocriminology is revolutionising our understanding of what makes humans commit crimes. Its core argument is that while the environment undoubtedly contributes to crime and violence, neurobiological factors also play a fundamental role. A word coined by the Canadian psychologist James Hilborn, neurocriminology is changing everything from prison sentences to the way we view rehabilitation. It also raises profound questions regarding the nature of free will and personal responsibility. If criminal behaviour is physically engrained somewhere in the inner recesses of the mind, can we really say that criminals are truly responsible for their actions? Perhaps it is like obesity, with some people's biology making them more likely to overeat and others less likely. In that case, perhaps a graded system of punishing criminals in which some receive shorter sentences despite having committed the same crime would be more appropriate than our current system.

Page was sentenced to life in prison without parole. In the years following Page's arrest, a British neuroscientist called Adrian Raine became interested in the science of criminology and decided to study Page's brain using functional brain imaging. He found reduced activity in the prefrontal cortex (which regulates impulse control) and increased activity in the amygdala (that regulates emotion). Raine can now use this distinctive brain pattern to predict with 70 per cent accuracy which criminals will reoffend following their release from prison.[19]

The discovery has already been put to use. In one study, Kent Kiehl, a neuroscientist at the Mind Research Network in Albuquerque, New Mexico, studied ninety-six male prisoners just before their release. He found that men who have lower activity in the anterior cingulate cortex, a small region at the front of the brain, had a 2.6-fold higher rate of rearrest for all crimes, and a 4.3-fold higher rate for nonviolent crimes.[20] In another study, researchers showed that if a released prisoner has a small amygdala, he or she is three times more likely to reoffend.[21] Remarkably, we may be able to use this information to make adjustments to our brains. Scientists at the University of Colorado, Boulder have created a brain implant as small as a human hair that can detect brain waves linked to violence, and alter a subject's brain chemistry accordingly.[22]

Discovering which brain regions are responsible for violence is a priority for neurocriminologists. In 2016, a ground-breaking study by Annegret Falkner and her colleagues at Princeton University showed that an area of the hypothalamus known as the ventromedial hypothalamus (VMH) becomes active during behaviour such as stalking, bullying and sexual aggression.[23] The hypothalamus is an ancient brain region that has been conserved throughout mammalian evolution. Located at the base of the brain, the hypothalamus controls sleep, hunger, thirst, sex, anger, hormone release and body temperature. With such a myriad of functions, pinning down its role in violence has been tricky. But strikingly, Falkner's study showed that the VHM becomes active in the moments leading up to an act of aggression, in those menacing seconds just when you feel your blood is beginning to boil. Now, researchers are investigating ways to turn the VHM on and off like a switch, arresting our violent impulses before they ever have a chance to harm others.

Science fiction enthusiasts will have already spotted where this is heading. In Phillip K. Dick's 1956 dystopian novella *Minority Report* (adapted into a 2002 film starring Tom Cruise), a specialised police department dubbed 'PreCrime' apprehends criminals before they commit their crimes using a group of psychics called 'precogs'. Swap psychics for a neural implant that monitors the VHM, and the novella may just have predicted the future. How scientists propose to switch the VHM on and off is a fascinating advance in its own right. The technique, known as optogenetics, uses light artificially to activate or deactivate groups of neurons. First thought up by Francis Crick in 1999 (discovering DNA clearly wasn't enough for him) and later perfected by Zhuo-Hua Pan at Wayne State University and Edward Boyden and his colleagues at Stanford University, optogenetics relies on injecting a light-sensitive protein (derived from a jellyfish) into a specific brain region. Before long, the neurons start producing the protein themselves, as if it had been there all along, allowing parts of the brain to be switched off with a beam of light.

Think about what happens when people get into a fight. First there's the verbal dispute and rising anger, which then quickly erupts into fisticuffs. Often, though certainly not always, a friend or passer-by sees what's happening and tries to break up the fight. Our social brains and evolved moral sense intervene for the benefit of our species. But there's a heavy cost: people get hurt, sometimes killed. Imagine if that friend or passer-by was a chip implanted in our heads; not there all the time, of course, just whenever we need. Ah, but what about self-defence, you might think. Yet even self-defence would be unnecessary because, if every brain had such a chip, violence towards another human would be literally unthinkable.

But there doesn't have to be a chip imposed from outside. We also know that contemplative practices like mindfulness and meditation can, quite literally, rewire the brain. 'You can use your mind to change the function and structure of your brain,' explains Dan Siegel, a psychiatrist and mindfulness expert at the University of California, Los Angeles, whose research has shown that mindfulness meditation stimulates the growth of integrative fibres in the brain. Siegel believes that through mindfulness-based empathy and compassion, 'we can evolve better brains.'[24] Perhaps, then, world peace also lies in a greater understanding of how our brain's neurophysiology affects our mood and sense of well-being.

Mindfulness meditation relies on paying close attention to the present moment. All thoughts of the past and future must be silenced, and the person must simply focus on the thoughts, feelings and bodily sensations she is experiencing right now. In addition to its burgeoning popularity among Westerners in recent years, mindfulness has gripped the medical world, with studies showing that it ameliorates stress, anxiety, depression and even drug addiction. In 2017, moreover, scientists discovered its true biological impact: it lowers inflammatory molecules and stress hormones by around 15 per cent. Now, mindfulness is recommended by the UK's NHS and other leading healthcare providers around the globe.

Thankfully, while human history appears to suggest that the brain is inherently predisposed to violence, new research indicates that humans are actually evolving more peaceful and cooperative minds. In 2016, José María Gómez at the University of Granada conducted the first thorough survey of violence in the mammal world, collecting data on more than a thousand species.[25] Believe it or not, in addition to starring in adverts for car insurance, the mammal most likely to be murdered by its own kind is the meerkat. It roams the Kalahari Desert in gangs of fifty, with 19.4 per cent meeting their end at the claws and teeth of another meerkat. When *Homo sapiens* first evolved, our murder rate was about 2 per cent, meaning one in fifty would have been murdered by other people. This is six times more lethally violent than the average mammal and is a pretty high percentage when you consider that the average for all the species Gómez and his team studied is 0.3.

The rates of lethal violence in human societies has risen and fallen throughout history. During the Palaeolithic period (the Stone Age, up to 2.4 million years ago), our rate was thought to be about 3.9 per cent. During the Medieval period, that rate rose to around 12 per cent, only then to fall precipitously in recent centuries to a rate lower than when we first evolved.

It's not entirely clear how the brain is becoming more peaceable as it evolves. In his famous essay *Leviathan* (1651), Thomas Hobbes argued that peace and social harmony are only achieved through a social contract established by a strong government. Left to their own devices, Hobbes believed, humans are selfish, violent creatures – 'solitary, poore, nasty, brutish' – who will exist in a state of nature and constantly seek to destroy one another. By organising themselves into states, humans create what he called an 'artificial person', a 'common power' or 'political community' that enforces and protects civil society. In one of his many proclamations, he asserts: 'a man [must] be willing when others are so too . . . to lay down his right to all things; and be contented with so much liberty against other men, as he would allow other men against himself.'[26] After witnessing the British Civil Wars, Hobbes was starkly aware of the consequences of state failure, and of how easily humans will embrace brutality.

Hobbes was definitely onto something. Human minds adapt and develop in response to the external environment, and in our time that environment is heavily defined by large states and the rule of secular law.

Your mind is changing. Or rather, your mind is changing still. The human brain is an engine of change. Its biological powers have been bestowed upon us by millions of years of evolution. But from its beginnings in chance and mishap, our brain's history has evolved into a tale shaped by us – by our societies, our discoveries and our uncompromising quest for knowledge. Nothing in our brain is fixed or permanent. In its endless cycles of change, in its susceptibility to nurture, and in its power to shape our world, our minds offer us limitless possibilities.

Acknowledgements

I'd like to start by thanking everyone who agreed to be interviewed for this book. Due to the devastating Covid-19 pandemic, their time was given to me over the phone or online, both forms of communication requiring tremendous patience and generosity on their part. I am also extremely grateful to the scientists and scholars quoted here for sharing their expertise and wisdom. Their passion for discovery is the lifeblood of popular science writing; any shortcomings in conveying the brilliance of their work are mine.

I owe enormous gratitude to my amazing editors, Kate Craigie, Georgina Laycock, and Candida Brazil, whose comments and revisions made this book significantly better – and whose enthusiasm and support continues to inspire me as a writer. Thanks also to the entire team at John Murray, who work tirelessly to make science accessible to the public. I would also like to thank my brilliant US editor, Ian Straus, and the entire team at Little, Brown for all their hard work across the pond. Carrie Plitt, my wonderful agent at Felicity Bryan Associates, deserves special thanks – your early encouragement and unceasing guidance turned this book from idea to reality.

An incalculable debt is owed to my parents, Marcella and Abol, who have sacrificed so much for their children. Last but not least, I am eternally grateful to my partner, Olivia. Your boundless love, amazing mind and endless humour have made me happy in ways I never thought possible.

Notes

Chapter 1: Building the Human Brain

1. M. Brunet, F. Guy, D. Pilbeam et al. (2002), 'A New Hominid from the Upper Miocene of Chad, Central Africa', *Nature*, 418 (6,894): 145–51.
2. D. Falk (2009), 'The Natural Endocast of Taung (Australopithecus africanus): Insights from the Unpublished Papers of Raymond Arthur Dart', *American Journal of Physical Anthropology*, 140, Suppl. 49: 49–65.
3. S. Neubauer, J. J. Hublinand and P. Gunz (2018), 'The Evolution of Modern Human Brain Shape', *Science Advances*, 4 (1): eaao5961.
4. D. C. Johanson and M. A. Edey (1981), *Lucy: The Beginnings of Humankind*, New York, NY: Simon & Schuster.
5. M. Florio, M. Albert, E Taverna et al. (2015), 'Human-Specific Gene ARHGAP11B Promotes Basal Progenitor Amplification and Neocortex Expansion', *Science*, 347 (6,229): 1,465–70.
6. M. Hawrylycz, J. A. Miller, V. Menon et al. (2015), 'Canonical Genetic Signatures of the Adult Human Brain', *Nature Neuroscience*, 18 (12): 1,832–44.
7. M. Lam, J. W Trampush, J. Yu et al. (2017), 'Large-Scale Cognitive GWAS Meta-Analysis Reveals Tissue-Specific Neural Expression and Potential Nootropic Drug Targets', *Cell Reports*, 21 (9): 2,597–613.
8. H. H. Stedman, B. W. Kozyak, A. Nelson et al. (2004), 'Myosin Gene Mutation Correlates with Anatomical Changes in The Human Lineage', *Nature*, 428 (6,981): 415–18.
9. E. Pennisi (2004), 'Human Evolution. The Primate Bite: Brawn Versus Brain?', *Science*, 303 (5,666): 1,957.
10. S. Pinker (1997), *How the Mind Works*, New York, NY: W. W. Norton.
11. S. M. Boback, C. L. Cox, B. D. Ott et al. (2007), 'Cooking and Grinding Reduces the Cost of Meat Digestion', *Comparative Biochemistry and Physiology A: Molecular Integrative Physiology*, 148 (3): 651–6.

12. K. L. Allen and R. F. Kay (2012), 'Dietary Quality and Encephalization in Platyrrhine Primates', *Proceedings of the Royal Society of London. Series B, Containing Papers of a Biological Character*, 279 (1,729): 715–21.
13. B. V. Peechakara and M. Gupta (2020), 'Vitamin B3', Treasure Island, FL: StatPearls.
14. R. W. Wrangham (2010), *Catching Fire: How Cooking Made Us Human*, Profile, (pp. 79–80).
15. Z. Molnar, G. J. Clowry, N. Sestan et al. (2019), 'New Insights into the Development of the Human Cerebral Cortex', *Journal of Anatomy*, 235 (3): 432–51.
16. S. Benito-Kwiecinski, S. L. Giandomenico, M. Sutcliffe et al. (2021), 'An Early Cell Shape Transition Drives Evolutionary Expansion of the Human Forebrain', *Cell*, 184 (8): 2,084–102.
17. G. Eliot (1871), *Middlemarch*, Penguin Classics, 2011, p. 766.
18. P. Brown, T. Sutikna, M. J. Morwood et al. (2004), 'A New Small-Bodied Hominin from the Late Pleistocene of Flores, Indonesia'. *Nature*, 431 (7,012): 1,055–61.
19. S. Marek, J. S. Siegel, E. M. Gordon et al. (2018), 'Spatial and Temporal Organization of the Individual Human Cerebellum', *Neuron*, 100 (4): 977–93, e977.
20. P. Marien, H. Ackermann, M. Adamaszek et al. (2014), 'Consensus Paper: Language and the Cerebellum: An Ongoing Enigma', *Cerebellum*, 13 (3): 386–410.
21. H. S. Ghosh (2019), 'Adult Neurogenesis and the Promise of Adult Neural Stem Cells', *Journal of Experimental Neuroscience*, 13: 1179069519856876.
22. L. Feuillet, H. Dufour and J. Pelletier (2007), 'Brain of a White-Collar Worker', *Lancet*, 370 (9,583): 262.
23. W. Lutz, and E. Kebede (2018), 'Education and Health: Redrawing the Preston Curve', *Population and Development Review*, 44 (2): 343–61.
24. The World Economic Forum's website is an abundant source of data on the case for gender equality.

Chapter 2: Inventing Emotion

1. J. S. Feinstein, R. Adolphs, A. R. Damasio and D. Tranel (2011), 'The Human Amygdala and the Induction and Experience of Fear', *Current Biology*, 21 (1): 34–8.
2. A. Anderson (2007), 'Feeling Emotional: The Amygdala Links Emotional

Perception and Experience', *Social Cognitive and Affective Neuroscience*, 2 (2): 71–2.

3. W. James (1884), 'What is an Emotion?', *Mind*, 9 (34): 189–90.
4. C. Darwin (1872), *Emotions in Man and Animals*, John Murray.
5. Ibid., p. 115.
6. A. Damasio (2006), *Descartes' Error*, Vintage, p. 45.
7. Interview with the author.
8. M. Beard (2014), *Laughter in Ancient Rome: On Joking, Tickling, and Cracking Up* (Sather Classical Lectures), Berkley, CA: University of California Press, p. 75.
9. C. Crivelli, J. A. Russell, S. Jarillo and J. M. Fernandez-Dols (2016), 'The Fear Gasping Face as a Threat Display in a Melanesian Society', *Proceedings of the National Academy of Sciences of the United States of America*, 113 (44): 12,403–7.
10. P. Ekman (2004), *Emotions Revealed: Understanding Faces and Feelings*, Weidenfeld & Nicolson.
11. L. Feldman Barrett (2017), *How Emotions Are Made: The Secret Life of the Brain*, Macmillan, p. 53.
12. A. Solomon (2002), *The Noonday Demon: An Anatomy of Depression*, Vintage, p. 401.
13. S. P. Barbic, Z. Durisko and P. W. Andrews (2014), 'Measuring the Bright Side of Being Blue: A New Tool for Assessing Analytical Rumination in Depression', *PLoS ONE*, 9 (11): e112077.
14. L. B. Alloy and L. Y. Abramson (1988), 'Depressive Realism: Four Theoretical Perspectives, in L. B. Alloy (ed.), *Cognitive Processes in Depression* (pp. 223–65), New York, NY: Guilford Press, p. 223.
15. J. Forgas (2007), 'When Sad is Better Than Happy: Negative Affect Can Improve the Quality and Effectiveness of Persuasive Messages and Social Influence Strategies', *Journal of Experimental Social Psychology*, 43 (4): 513–28.
16. J. Forgas, L. Goldenberg and C. Unkelbach (2009), 'Can Bad Weather Improve Your Memory? An Unobtrusive Field Study of Natural Mood Effects on Real-Life Memory', *Journal of Experimental Social Psychology*, 45 (1): 254–7.
17. J. Rottenberg (2014), *The Depths: The Evolutionary Origins of the Depression Epidemic*, New York, NY: Basic Books, p. 24.
18. From a talk given by Ed Bullmore, London, September 2017, to coincide with the Academy of Medical Sciences FORUM annual lecture.
19. E. Bullmore (2019), *The Inflamed Mind: A Radical New Approach to Depression*, Short Books, p. 19.

20. B. H. Hidaka (2012), 'Depression as a Disease of Modernity: Explanations for Increasing Prevalence', *Journal of Affective Disorders*, 140 (3): 205–14.
21. Quoted in N. Casey (2002), *Unholy Ghost: Writers on Depression*, New York, NY: William Morrow.
22. H. Rose Markus and B. Schwartz (2010), 'Does Choice Mean Freedom and Well-Being? *Journal of Consumer Research*, 37 (2): 344–55.
23. J. L. Frijling (2017), 'Preventing PTSD with Oxytocin: Effects of Oxytocin Administration on Fear Neurocircuitry and PTSD Symptom Development in Recently Trauma-Exposed Individuals', *European Journal of Psychotraumatology*, 8 (1): 1302652.
24. D. M Cochran, D. Fallon, M. Hill and J. A. Frazier (2013), 'The Role of Oxytocin in Psychiatric Disorders: A Review of Biological and Therapeutic Research Findings', *Harvard Review of Psychiatry*, 21 (5): 219–47.
25. I. Schneiderman, O. Zagoory-Sharon, J. F. Leckman and R. Feldman (2012), 'Oxytocin During the Initial Stages of Romantic Attachment: Relations to Couples' Interactive Reciprocity', *Psychoneuroendocrinology*, 37 (8): 1,277–85.
26. D. Scheele, N. Striepens, O. Güntürkün et al. (2012), 'Oxytocin Modulates Social Distance Between Males and Females', *Journal of Neuroscience*, 32 (46): 16,074–9.
27. M. Lek, K. J. Karczewski, E. V. Minikel et al. (2016), 'Analysis of Protein-Coding Genetic Variation in 60,706 Humans', *Nature*, 536 (7,616): 285–91.
28. T. Doi (1971, repr. 2002), *The Anatomy of Dependence*, New York, NY: Kodansha America.

Chapter 3: Our Social Brains

1. S. R. H. Beach, G. H. Brody, T. D. Gunter et al. (2010), 'Child Maltreatment Moderates the Association of MAOA with Symptoms of Depression and Antisocial Personality Disorder', *Journal of Family Psychology*, 24 (1): 12–20.
2. M. R. Rautiainen, T. E. Paunio, E. Repo-Tiihonen et al. (2016), 'Genome-Wide Association Study of Antisocial Personality Disorder', *Translational Psychiatry*, 6 (9): e883.
3. T. D. White, B. Asfaw, Y. Beyene et al. (2009), 'Ardipithecus ramidus and the Paleobiology of Early Hominids', *Science*, 326 (5,949): 75–86.

4. M. Lieberman (2015), *Social: Why Our Brains Are Wired to Connect*, Oxford: Oxford University Press, p. 10.
5. D. Reiss, L. Leve and J. Neiderhiser (2013), 'How Genes and the Social Environment Moderate Each Other', *American Journal of Public Health*, 103: S111–21.
6. S. B. Hrdy (2011), *Mothers and Others: The Evolutionary Origins of Mutual Understanding*, Cambridge, MA: Belknap Press, p. 3.
7. C. Darwin (1871), *The Descent of Man, and Selection in Relation to Sex*, John Murray.
8. V. Horner, J. D. Carter, M, Suchak and F. B. de Waal (2011), 'Spontaneous Prosocial Choice by Chimpanzees', *Proceedings of the National Academy of Sciences of the United States of America*, 108 (33): 13,847–51.
9. A. Shahaeian, C. C. Peterson, V. Slaughter and H. M. Wellman (2011), 'Culture and the Sequence of Steps in Theory of Mind Development', *Developmental Psychology*, 47 (5): 1,239–47.
10. A. Mayer and B. E. Träuble (2013), 'Synchrony in the Onset of Mental State Understanding Across Cultures? A Study Among Children in Samoa', *International Journal of Behavioural Development*, 37: 21–8.
11. A. Mizokawa and S. Lecce (2016), 'Sensitivity to Criticism and Theory of Mind: A Cross Cultural Study on Japanese and Italian Children', *European Journal of Developmental Psychology*, 14 (2): 159–71.
12. J. Henrich and J. Broesch (2011), 'On the Nature of Cultural Transmission Networks: Evidence from Fijian Villages for Adaptive Learning Biases', *Philosophical Transactions of the Royal Society of London: B. Biological Sciences*, 366 (1,567): 1,139–48.
13. R. B. Lee (1969), 'Eating Christmas in the Kalahari', http://people.morrisville.edu/~reymers/readings/ANTH101/EatingChristmas-Lee.pdf
14. M. Sheskin (2018), 'The Inequality Delusion: Why We've Got the Wealth Gap All Wrong', www.newscientist.com/article/mg23731710-300-the-inequality-delusion-why-weve-got-the-wealth-gap-all-wrong/
15. P. Singer (2011), *The Expanding Circle: Ethics, Evolution, and Moral Progress*, Princeton, NJ: Princeton University Press.
16. R. I. M. Dunbar (1992), 'Neocortex Size as a Constraint on Group Size in Primates', *Journal of Human Evolution*, 22 (6): 469–93.
17. M. E. Brashears (2011), 'Small Networks and High Isolation? A Reexamination of American Discussion Networks', *Social Networks*, 33 (4): 331–41.
18. R. S. Burt (2004), 'Structural Holes and Good Ideas', *American Journal of Sociology*, 110 (2): 349–99.

19. M. B. O'Donnell, J. B. Bayer, C. N. Cascio and E. B. Falk (2017), 'Neural Bases of Recommendations Differ According to Social Network Structure', *Social Cognitive and Affective Neuroscience*, 12 (1): 61–9.
20. J. B. Silk (2007), 'Social Components of Fitness in Primate Groups', *Science*, 317 (5,843): 1,347–51.
21. D. C. Johanson (2004), 'Lucy, Thirty Years Later: An Expanded View of Australopithecus afarensis', *Journal of Anthropological Research*, 60 (4): 465–86.
22. E. Nelson, C. Rolian, L. Cashmore and S. Shultz (2011), 'Digit Ratios Predict Polygyny in Early Apes, Ardipithecus, Neanderthals and Early Modern Humans but Not in Australopithecus', *Proceedings of the Royal Society of London. Series B, Containing Papers of a Biological Character*, 278 (1,711): 1,556–63.
23. D. Labuda, J.-F. Lefebvre, P. Nadeau and M. H. Roy-Gagnon (2010), 'Female-to-Male Breeding Ratio in Modern Humans: An Analysis Based on Historical Recombinations', *American Journal of Human Genetics*, 86 (3): 353–63.
24. P. E. Komers and P. N. M. Brotherton (1997), 'Female Space Use is the Best Predictor of Monogamy in Mammals', *Proceedings of the Royal Society of London. Series B*, 264: 1,261–70.
25. L. Williamson (1978), 'Infanticide: An Anthropological Analysis', in M. Kohl (ed.), *Infanticide and the Value of Life* (pp. 61–75:), Lanham, MD: Prometheus Books
26. R. J. Sampson, J. H. Laub and C. Wimer (2006), 'Does Marriage Reduce Crime? A Counterfactual Approach to Within-Individual Causal Effects', *Criminology*, 44 (3): 465–508.
27. M. Huck, E. Fernandez-Duque, P. Babb and T. Schurr (2014), 'Correlates of Genetic Monogamy in Socially Monogamous Mammals: Insights from Azara's Owl Monkeys', *Proceedings of the Royal Society of London. Series B, Containing Papers of a Biological Character*, 281 (1,782): 20140195.

Chapter 4: The Genesis of Memory

1. Quoted in D. Draaisma, and P. Vincent (2000), *Metaphors of Memory: A History of Ideas About the Mind*, Cambridge: Cambridge University Press, p. 27.
2. E. A. Murray, S. P. Wise and K. S. Graham (2017), *The Evolution of*

Memory Systems: Ancestors, Anatomy, and Adaptations (Oxford Psychology Series), Oxford: Oxford University Press.

3. H. Bunn (2012), talk given at the European Society for the Study of Human Evolution.
4. J. Manning (2016), 'Come to Think of It . . . or Not: Dartmouth Study Shows How Memories Can be Intentionally Forgotten', www.dartmouth.edu/press-releases/memories-can-be-intentionally-forgotten-050516.html
5. R. Shadmehr, J. Brand and S. Corkin (1998), 'Time-Dependent Motor Memory Processes in Amnesic Subjects', *Journal of Neurophysiology*, 80 (3): 1,590–7.
6. D, Kahneman (2012), *Thinking, Fast and Slow*, Penguin.
7. J. Holmes, K. A. Hilton, M. Place et al. (2014), 'Children with Low Working Memory and Children with ADHD: Same or Different?', *Frontiers in Human Neuroscience*, 8: 976.
8. R. Martinussen, J. Hayden, S. Hogg-Johnson and R. Tannock (2005), 'A Meta-Analysis of Working Memory Impairments in Children with Attention-Deficit/Hyperactivity Disorder', *Journal of the American Academy of Child and Adolescent Psychiatry*, 44 (4): 377–84.
9. N. V. Kukushkin and T. J. Carew (2017), 'Memory Takes Time', *Neuron*, 95 (2): 259–79.
10. M. Proust (1913), *In Search of Lost Time*, Penguin Modern Classics, 2003, p. 47.
11. N. S. Jacobs, T. A. Allen, N. Nguyen and N. J. Fortin (2013), 'Critical Role of the Hippocampus in Memory for Elapsed Time', *Journal of Neuroscience*, 33 (34): 13,888–93.
12. A. J. Schwartz, A. Boduroglu and A. H. Gutchess (2014), 'Cross-Cultural Differences in Categorical Memory Errors', *Cognitive Science*, 38 (5): 997–1,007.
13. Quoted in J. K. Foster (2008), *Memory: A Very Short Introduction*, Oxford: Oxford University Press, p. 15.
14. M. Edelson, T. Sharot, R. J. Dolan and Y. Dudai (2011), 'Following the Crowd: Brain Substrates of Long-Term Memory Conformity', *Science*, 333 (6,038): 108–11.
15. M. B. Reysen (2007), 'The Effects of Social Pressure on False Memories', *Memory and Cognition*, 35 (1): 59–65.
16. W. Hirst, J. K. Yamashiro and A. Coman (2018), 'Collective Memory from a Psychological Perspective', *Trends in Cognitive Science*, 22 (5): 438–51.

Chapter 5: The Truth About Intelligence

1. J. Newman (1956), *The World of Mathematics*, vol. 1, New York, NY: Dover, p. 369.
2. M. R. Murty and V. K. Murty (2013), *The Mathematical Legacy of Srinivasa Ramanujan*, New York, NY: Springer, p. 5.
3. L. Gabora, and A. Russon (2011), 'The Evolution of Intelligence', in *The Cambridge Handbook of Intelligence* (pp. 328–42), Cambridge: Cambridge University Press, p. 332.
4. D. C. Dennett (2018), *From Bacteria to Bach and Back: The Evolution of Minds*, Penguin, p. 196.
5. G. E. Hinton and S. J. Nolan (1987), 'How Learning Can Guide Evolution', *Complex Systems*, 1: 495–502.
6. E. Guerra-Doce (2015), 'Psychoactive Substances in Prehistoric Times: Examining the Archaeological Evidence', *Journal of Archaeology, Consciousness and Culture*, 8 (1): 91–112.
7. R. Power, D. Salazar-Garcia, L. Straus et al. (2015), 'Microremains from El Mirón Cave Human Dental Calculus Suggest a Mixed Plant–Animal Subsistence Economy During the Magdalenian in Northern Iberia', *Journal of Archaeological Science*, 60: 39–46.
8. H. S. Kühl, A. K. Kalan, M. Arandjelovic et al. (2016), 'Chimpanzee Accumulative Stone Throwing', *Scientific Reports*, 6: 22219.
9. D. A. Leavens, K. A. Bard and W. D. Hopkins (2019), 'The Mismeasure of Ape Social Cognition', *Animal Cognition*, 22 (4): 487–504.
10. F. De Waal (2017), *Are We Smart Enough to Know How Smart Animals Are?* Granta, p. 157.
11. A. Binet and T. Simon (1905), 'New Methods for the Diagnosis of the Intellectual Level of Subnormals', in A Binet and T. Simon, *The Development of Intelligence in Children: The Binet-Simon Scale* (pp. 37–90), trans. E. S. Kite, Baltimore, MD: Williams & Wilkins, 1916.
12. G. Stoddard (1943), *The Meaning of Intelligence*, Macmillan.
13. D. Wechsler (1944), *The Measurement of Adult Intelligence*, Baltimore, MD: Williams & Wilkins.
14. L. G. Humphreys (1979), 'The Construct of General Intelligence', *Intelligence*, 3 (2): 105–20.
15. L. Gottfredson (1998), 'The General Intelligence Factor. Scientific', *American Presents*, 9 (4): 24–9.
16. S. Aaronson (2017), 'Also Against Individual IQ Worries', blog, 1 October 2017, https://www.scottaaronson.com/blog/?p=3473

17. R. Sternberg interview for *Readers Digest*, https://www.rd.com/culture/practical-intelligence/
18. B. S. Boxer Wachler (2017), *Perceptual Intelligence: The Brain's Secret to Seeing Past Illusion, Misperception, and Self-Deception*, Novato, CA: New World Library.
19. R. A. Barton and C. Venditti (2013), 'Human Frontal Lobes Are Not Relatively Large', *Proceedings of the National Academy of Sciences of the United States of America*, 110 (22): 9,001–6.
20. S. B. Kaufman (2016), talk given at the Bay Area Discovery Museum's Creativity Forum, Fairmont San Francisco Hotel, San Francisco, CA.
21. A. J. Horner, J. A. Bisby, E. Zotow et al. (2016), 'Grid-Like Processing of Imagined Navigation', *Current Biology*, 26 (6): 842–7.
22. T. D. Wilson, D. A. Reinhard, E. C. Westgate et al. (2014), 'Just Think: The Challenges of the Disengaged Mind', *Science*, 345 (6,192): 75–7.
23. N. H. Immordino-Yang, J. A Christodoulou and V. Singh (2012), 'Rest is Not Idleness: Implications of the Brain's Default Mode for Human Development and Education', *Perspectives on Psychological Science*, 7 (4): 352–64.
24. R. J. Sternberg and S. B. Kaufman (2011), *The Cambridge Handbook of Intelligence*, Cambridge: Cambridge University Press.
25. S. M. McKinney, M. Sieniek, V. Godbole et al. (2020), 'International Evaluation of an AI System for Breast Cancer Screening', *Nature*, 577 (7,788): 89–94.
26. J. Flynn (2013), 'The Flynn Effect and Flynn's Paradox', *Intelligence*, 41: 851–7.
27. L. Albantakis, A. Hintze, C. Koch et al. (2014), 'Evolution of Integrated Causal Structures in Animats Exposed to Environments of Increasing Complexity', *PLoS Computational Biology*, 10 (12): e1003966.
28. S. Ramsden, F. M. Richardson, G. Josse et al. (2011), 'Verbal and Non-Verbal Intelligence Changes in the Teenage Brain', *Nature*, 479 (7,371): 113–16.
29. Report by D. Shukman, BBC, 19 October 2011.
30. R. Rosenthal and L. Jacobson (1968), 'Pygmalion in the Classroom', *Urban Review*, 3 (1): 16–20.
31. J. Protzko (2016), 'Does the Raising IQ-Raising g Distinction Explain the Fadeout Effect?' *Intelligence*, 56: 65–71.

Chapter 6: Creating Language

1. D. Everett (2018), interview with Nicola Davis for the *Guardian*: '*Homo erectus* May Have Been a Sailor – and Able to Speak', 20 February 2018, www.theguardian.com/science/2018/feb/20/homo-erectus-may-have-been-a-sailor-and-able-to-speak
2. J. Monod (1971), *Chance and Necessity: An Essay on the Natural Philosophy of Modern Biology*, Vintage, p. 133.
3. C. Heyes (2018), *Cognitive Gadgets: The Cultural Evolution of Thinking*, Cambridge, MA: Belknap Press.
4. M. Ruhlen (2011), interviewed for Life's Little Mysteries, quoted at LiveScience, www.livescience.com/16541-original-human-language-yoda-sounded.html
5. D. Smith, P. Schlaepfer, K. Major et al. (2017), 'Cooperation and the Evolution of Hunter-Gatherer Storytelling', *Nature Communications*, 8 (1): 1,853.
6. J. Odling-Smee and K. N. Laland (2009), 'Cultural Niche Construction: Evolution's Cradle of Language', in R. Botha and C. Knight (eds), *The Prehistory of Language*, New York, NY: Oxford University Press, p. 99.
7. J. D. Yeatman, R. F. Dougherty, M. Ben-Shachar and B. A. Wandell (2012), 'Development of White Matter and Reading Skills', *Proceedings of the National Academy of Sciences of the United States of America*, 109 (44): E3045–53.
8. I. Maddieson (2015), talk given at the 170th Meeting of the Acoustical Society of America (ASA), quoted in E. Underwood, 'Human Language May Be Shaped by Climate and Terrain', www.sciencemag.org/news/2015/11/human-language-may-be-shaped-climate-and-terrain
9. S. Curtiss (2016), quoted by Rory Carroll in the *Guardian,* 'Starved, Tortured, Forgotten: Genie, the Feral Child who Left a Mark on Researchers', 14 July 2016, www.theguardian.com/society/2016/jul/14/genie-feral-child-los-angeles-researchers
10. M. Pagel (2017), 'Darwinian Perspectives on the Evolution of Human Languages', *Psychonomic Bulletin and Review*, 24 (1): 151–7.
11. M. G. Newberry, C. A. Ahern, R. Clark and J. B. Plotkin (2017). 'Detecting Evolutionary Forces in Language Change', *Nature*, 551 (7,679): 223–6.
12. A. D. Friederici (2017), 'Language in Our Brain: The Origins of a Uniquely Human Capacity', Cambridge, MA: MIT Press, p. ix.
13. D. A. Moses, M. K. Leonard, J. G. Makin and E. F. Chang (2019), 'Real-

Time Decoding of Question-and-Answer Speech Dialogue Using Human Cortical Activity', *Nature Communications*, 10 (1): 3,096.

14. W. T. Fitch, B. de Boer, N. Mathur and A. A. Ghazanfar (2016), 'Monkey Vocal Tracts Are Speech-Ready', *Science Advances*, 2 (12): e1600723.
15. W. T. Fitch (2016), interviewed by Carl Zimmer for the *New York Times*, 'Monkeys Could Talk, but They Don't Have the Brains for It', 9 December 2016, www.nytimes.com/2016/12/09/science/monkeys-speech.html
16. P. Lieberman (2017), 'Comment on "Monkey Vocal Tracts are Speech-ready"', *Science Advances*, 3 (7): e1700442.
17. J. Arac (2002), 'Anglo-Globalism?', *New Left Review*, 16, July/August, https://newleftreview.org/issues/II16/articles/jonathan-arac-anglo-globalism
18. N. Ostler, (2018), 'Have We Reached Peak English in the World?', *Guardian*, 27 February 2018, www.theguardian.com/commentisfree/2018/feb/27/reached-peak-english-britain-china

Chapter 7: The Illusion of Consciousness

1. T. Nagel (1974), 'What Is It Like to Be a Bat?', *Philosophical Review*, 83 (4): p. 436.
2. For a comprehensive account of ideas about consciousness, see the wonderful S. Blakemore and E. T. Trosciannoko (2018), *Consciousness: An Introduction*, Routledge.
3. N. Charter (2018), *The Mind is Flat: The Illusion of Mental Depth and the Improvised Mind*, Allen Lane, p. 186.
4. D. Papineau (2003), 'Confusions About Consciousness', *Richmond Journal of Philosophy*, 5.
5. D. Wolman (2012), 'The Split Brain: A Tale of Two Halves', *Nature*, 483: 260–3.
6. S. Greenfield (2015), *A Day in the Life of the Brain: The Neuroscience of Consciousness from Dawn Till Dusk*, Penguin.
7. N. Humphrey (1992), *A History of the Mind: Evolution and the Birth of Consciousness*, Göttingen: Copernicus.
8. M. S. A. Graziano and T. W. Webb (2014), 'A Mechanistic Theory of Consciousness', *International Journal of Machine Consciousness*, 6: 163–76.
9. J. Joyce (1922), *Ulysses*, Penguin Classics, 2015, p. 204.
10. D. Hume (1739), *A Treatise of Human Nature: Being an Attempt to Introduce*

the Experimental Method of Reasoning into Moral Subjects, Penguin Classics, 1985, p. 252.

11. T. Reid (1785), *Essays on the Intellectual Powers of Man*, Edinburgh: John Bell.
12. G. Strawson (1997), 'The Self', *Journal of Consciousness Studies*, 4 (5–6): 424, in S. Blakemore and E. Troscianoko (2018), *Consciousness: An Introduction*, Routledge, p. 453.
13. T. Metzinger (2009), *The Ego Tunnel: The Science of the Mind and the Myth of the Self*, New York, NY: Basics Books, p. 115.
14. L. Chang, S. Zhang, M. M. Poo and N. Gong (2017), 'Spontaneous Expression of Mirror Self-Recognition in Monkeys After Learning Precise Visual-Proprioceptive Association for Mirror Images', *Proceedings of the National Academy of Sciences of the United States of America*, 114 (12): 3,258–63.
15. A. W. Huttunen, G. K. Adams and M. L. Platt. (2018), 'Can Self-Awareness Be Taught? Monkeys Pass the Mirror Test – Again', *Proceedings of the National Academy of Sciences of the United States of America*, 114 (13): 3,281–3.
16. For an excellent overview of narcissistic personality disorder, see the Mayo Clinic's website, www.mayoclinic.org/diseases-conditions/narcissistic-personality-disorder/symptoms-causes/syc-20366662
17. J.-J. Rousseau (1782), *The Confessions of Jean-Jacques Rousseau*, Penguin Classics, 1973.
18. B. Libet, C. A. Gleason, E. W. Wright and D. K. Pearl (1983), 'Time of Conscious Intention to Act in Relation to Onset of Cerebral Activity (Readiness-Potential): The Unconscious Initiation of a Freely Voluntary Act', *Brain*, 106 (pt 3): 623–42.
19. A. Harris (2019), *Conscious: A Brief Guide to the Fundamental Mystery of the Mind*, Harper, p. 34.
20. C. S. Soon, M. Brass, H. J. Heinze and J. D. Haynes (2008), 'Unconscious Determinants of Free Decisions in the Human Brain', *Nature Neuroscience*, 11 (5): 543–5.
21. Z. Boron (2014), 'How Powerful is a Mind? Supercomputer Takes 40 Minutes to Map 1 Second of Brain Activity', *Independent*, 14 January 2014, www.independent.co.uk/life-style/gadgets-and-tech/how-powerful-is-a-mind-supercomputer-takes-40-minutes-to-map-1-second-of-brain-activity-9059225.html
22. S. Cave (2016), 'There's No Such Thing as Free Will – But We're Better Off Believing in it Anyway', *Atlantic*, June 2016, www.theatlantic.com/magazine/archive/2016/06/theres-no-such-thing-as-free-will/480750/
23. D. M. Wegner and T. Wheatley (1999), 'Apparent Mental Causation: Sources of the Experience of Will', *American Psychologist*, 54 (7): 480–92.

Chapter 8: Different Minds

1. U. Frit (2008), *Autism: A Very Short Introduction*, Oxford: Oxford University Press, pp. 61–4.
2. K. Yoshida, Y. Go, I. Kushima et al. (2016), 'Single-Neuron and Genetic Correlates of Autistic Behavior in Macaque', *Science Advances*, 2 (9): e1600558.
3. A. Nadler, C. F. Camerer, D. T. Zava et al. (2019), 'Does Testosterone Impair Men's Cognitive Empathy? Evidence from Two Large-Scale Randomized Controlled Trials', *Proceedings of the Royal Society B: Biological Sciences*, 286 (1,910).
4. P. Vitebsky (2005), *Reindeer People: Living with Animals and Spirits in Siberia*, Harper Perennial.
5. This and the following quotes from V. Fenn (2018), 'These 10 Artists Prove Autism is No Barrier to Creativity', *Metro*, 21 April 2018, https://metro.co.uk/2018/04/21/these-10-artists-prove-autism-is-no-barrier-to-creativity-7446895/
6. L. Mottron, M. Dawson, I. Soulières et al. (2006), 'Enhanced Perceptual Functioning in Autism: An Update, and Eight Principles of Autistic Perception', *Journal of Autism and Developmental Disorders*, 36 (1): 27–43.
7. M. Haddon (2004), *The Curious Incident of the Dog in the Night-Time*, Vintage, p. 7.
8. T. Langdell (1978), 'Recognition of Faces: An Approach to the Study of Autism', *Journal of Child Psychology and Psychiatry*, 19 (3): 255–68.
9. S. Fishwick (2016), 'Spooky London: How Britain's Spy Agencies Are Using New and Unexpected Methods to Recruit the Next Generation', *Evening Standard*, 25 February 2016, www.standard.co.uk/lifestyle/esmagazine/spooky-london-the-new-methods-of-britains-spies-a3189396.html
10. C. Best, S. Arora, F. Porter and M. Doherty (2015), 'The Relationship Between Subthreshold Autistic Traits, Ambiguous Figure Perception and Divergent Thinking', *Journal of Autism and Developmental Disorders*, 45: 4,064–73.
11. J. L. Aragón, G. G. Naumis, M. Bai et al. (2008), 'Turbulent Luminance in Impassioned van Gogh Paintings', *Journal of Mathematical Imaging and Vision*, 30: 275–83.
12. M. Solomon, A. M. Iosif, V. P. Reinhardt et al. (2017), 'What Will My Child's Future Hold? Phenotypes of Intellectual Development in 2–8-Year-Olds with Autism Spectrum Disorder', *Autism Research*, 11 (1): 121–32.

13. Quoted in K. Moisse (2017), 'Many Children with Autism Get Significantly Smarter Over Time', *Spectrum News*, www.spectrumnews.org/news/many-children-autism-get-significantly-smarter-time/
14. 'Muskie' (2002), Institute for the Study of the Neurotypical, http://erikengdahl.se/autism/isnt/
15. J. Sinclair (1993), 'Don't Mourn for Us', www.autreat.com/dont_mourn.html
16. J. Singer (2017), *Neurodiversity: The Birth of an Idea*, Judy Singer, p. 20.
17. T. Grandin and R. Panek (2014), *The Autistic Brain*, Rider, p. 102.
18. S. Baron-Cohen, S. Wheelwright, R. Skinner et al. (2001), 'The Autism-Spectrum Quotient (AQ): Evidence from Asperger Syndrome/High-Functioning Autism, Males and Females, Scientists and Mathematicians', *Journal of Autism and Developmental Disorders*, 31 (1): 5–17.
19. L. Lundqvist and H. Lindner (2017), 'Is the Autism-Spectrum Quotient a Valid Measure of Traits Associated with the Autism Spectrum? A Rasch Validation in Adults With and Without Autism Spectrum Disorders', *Journal of Autism and Developmental Disorders*, 47 (7): 2,080–91.
20. Based on interviews with the author. Susan, Adam, Gwenn, Lee and Ramila are pseudonyms.
21. J. P. Allen, M. M. Schad, B. Oudekerk and J. Chango (2014), 'What Ever Happened to the "Cool" Kids? Long-Term Sequelae of Early Adolescent Pseudomature Behavior', *Child Development*, 85 (5): 1,866–80.
22. R. Zamzow (2019), 'Rethinking Repetitive Behaviours in Autism', *Spectrum News*, www.spectrumnews.org/features/deep-dive/rethinking-repetitive-behaviors-in-autism/
23. The poll of more than 300 HR professionals was performed in 2018 by the Chartered Institute of Personnel and Development.
24. H. Devlin (2018), 'Thousands of Autistic Girls and Women "Going Undiagnosed" Due to Gender Bias', *Guardian*, 14 September 2018, www.theguardian.com/society/2018/sep/14/thousands-of-autistic-girls-and-women-going-undiagnosed-due-to-gender-bias

Chapter 9: The iBrain

1. M. A. Lancaster, M. Renner, C. A. Martin et al. (2013), 'Cerebral Organoids Model Human Brain Development and Microcephaly', *Nature*, 501 (7,467): 373–9.
2. D. E. Redmond, Jr, S. Weiss, J. D. Elsworth et al. (2010), 'Cellular Repair

in the Parkinsonian Nonhuman Primate Brain', *Rejuvenation Research*, 13 (2–3): 188–94.

3. Quoted in K. Brown (2016), interview with Jürgen Knoblich, Company of Biologists, www.youtube.com/watch?v=hhmh8YKDwyo
4. J. Gil-Ranedo, E. Gonzaga, K. J. Jaworek et al. (2019), 'STRIPAK Members Orchestrate Hippo and Insulin Receptor Signaling to Promote Neural Stem Cell Reactivation', *Cell Reports*, 27 (10): 2,921–33 e2925.
5. S. Mlot (2018), 'World Go Champion Hopes to Reclaim Title in AI Rematch', www.geek.com/tech/world-go-champ-hopes-to-reclaim-title-in-ai-rematch-1726687/
6. S. Dehaene, H. Lau and S. Kouider (2017), 'What is Consciousness, and Could Machines Have It?', *Science*, 358 (6,362): 486–92.
7. J. Bauby (2008), *The Diving-Bell and the Butterfly*, HarperCollins, p. 47.
8. Quoted in P. Harris (2011), 'BrainGate Gives Paralysed the Power of Mind Control', *Guardian*, 17 April 2011, www.theguardian.com/science/2011/apr/17/brain-implant-paralysis-movement
9. D. J. McFarland, J. Daly, C. Boulay and M. Parvaz (2017), 'Therapeutic Applications of BCI Technologies', *Brain Computer Interfaces* (Abingdon), 47 (1–2): 37–52.
10. R. Juskalian (2020), 'A New Implant for Blind People Jacks Directly into the Brain', *MIT Technology Review*, www.technologyreview.com/s/615148/a-new-implant-for-blind-people-jacks-directly-into-the-brain/
11. M. A. Cervera, S. R. Soekadar, J. Ushiba et al. (2018), 'Brain-Computer Interfaces for Post-Stroke Motor Rehabilitation: A Meta-Analysis', *Annals of Clinical and Translational Neurology*, 5 (5): 651–63.
12. M. Z. Donahue (2017), 'How a Color-Blind Artist Became the World's First Cyborg', *National Geographic*, www.nationalgeographic.com/news/2017/04/worlds-first-cyborg-human-evolution-science/
13. M. E. Clynes and N. S. Kline (1960), 'Cyborgs and Space' *Astronautics*, September 1960: 26–7, 74–6.
14. K. Swisher and L. Goode (2017), 'Recode: Too Embarrassed to Ask', *Vox*, www.vox.com/2017/3/30/15130136/transcript-mary-lou-jepsen-one-laptop-per-child-too-embarrassed-to-ask-live-sxsw
15. G. Shen, T. Horikawa, K. Majima and Y. Kamitani (2019), 'Deep Image Reconstruction from Human Brain Activity', *PLoS Computational Biology*, 15 (1): e1006633.
16. R. P. Rao A. Stocco, M. Bryan et al. (2014), 'A Direct Brain-to-Brain Interface in Humans', *PLoS ONE*, 9 (11): e111332.
17. The Royal Society (2019), *iHuman: Blurring Lines Between Mind and Machine*, September 2019, Royal Society, p. 67.

18. M. Mosley (2015), 'Are Murderers Born or Made?', www.bbc.co.uk/news/magazine-31714853
19. A. Raine (2014), *The Anatomy of Violence: The Biological Roots of Crime*, Penguin.
20. E. Aharoni, G. M. Vincent, C. L. Harenski et al. (2013), 'Neuroprediction of Future Rearrest', *Proceedings of the National Academy of Sciences of the United States of America*, 110 (15): 6,223–8.
21. D. A. Pardini, A. Raine, K. Erickson and R. Loeber (2014), 'Lower Amygdala Volume in Men is Associated with Childhood Aggression, Early Psychopathic Traits, and Future Violence', *Biological Psychiatry*, 75 (1):73–80.
22. R. Qazi, A. M. Gomez, D. C. Castro et al. (2019), 'Wireless Optofluidic Brain Probes for Chronic Neuropharmacology and Photostimulation', *Nature Biomedical Engineering*, 3 (8): 655–69.
23. A. L. Falkner, L. Grosenick, T. J. Davidson et al. (2016), 'Hypothalamic Control of Male Aggression-Seeking Behavior', *Nature Neuroscience*, 19 (4): 596–604.
24. D. Siegel (2021), 'Brain Insights and Well-Being', www.drdansiegel.com/brain-insights-and-well-being-2/
25. J. M. Gómez, M. Verdu, A. Gonzalez-Megias and M. Mendez (2016), 'The Phylogenetic Roots of Human Lethal Violence', *Nature*, 538 (7,624): 233–7.
26. T. Hobbes (1651), *Leviathan*, Penguin Classics, 2016.

Bibliography

Aaronson, S. (2017), 'Also Against Individual IQ Worries', blog, 1 October 2017, https://www.scottaaronson.com/blog/?p=3473

Abe, K. and D. Watanabe (2011), 'Songbirds Possess the Spontaneous Ability to Discriminate Syntactic Rules', *Nature Neuroscience* 14 (8): 1,067–74

Abutalebi, J. (2008), 'Neural Aspects of Second Language Representation and Language Control', *Acta Psychologica*, 128 (3): 466–78

Adolphs, R. (2015), 'The Unsolved Problems of Neuroscience', *Trends in Cognitive Sciences*, 19 (4): 173–5

Adolphs, R., H. Damasio, D. Tranel et al. (2000), 'A Role for Somatosensory Cortices in the Visual Recognition of Emotion as Revealed by Three-Dimensional Lesion Mapping', *Journal of Neuroscience*, 20: 2,683–90

Adolphs, R., F. Gosselin, T. Buchanan et al. (2005), 'A Mechanism for Impaired Fear Recognition in Amygdala Damage', *Nature*, 433: 68–72

Aguzzi, A., B. A. Barres and M. L. Bennett (2013), 'Microglia: Scapegoat, Saboteur, Or Something Else?', *Science*, 339 (6,116): 156–61

Aharoni, E., G. M. Vincent, C. L. Harenski et al. (2013), 'Neuroprediction of Future Rearrest', *Proceedings of the National Academy of Sciences of the United States of America*, 110 (15): 6,223–8

Alaerts, K., K. Nayar, C. Kelly et al. (2015), 'Age-Related Changes in Intrinsic Function of the Superior Temporal Sulcus in Autism Spectrum Disorders', *Social Cognitive and Affective Neuroscience*, 10 (10): 1,413–23

Albantakis, L., A. Hintze, C. Koch et al. (2014), 'Evolution of Integrated Causal Structures in Animats Exposed to Environments of Increasing Complexity', *PLoS Computational Biology*, 10 (12): e1003966

Alberini, C. M. (2009), 'Transcription Factors in Long-Term Memory and Synaptic Plasticity', *Physiological Reviews*, 89: 121–45

Aleman, A. (2014), *Our Ageing Brain*, trans. A. Mills, Melbourne and London: Scribe

Alkire, M., A. G. Hudetz and G. Tononi (2008), 'Consciousness and Anesthesia', *Science*, 322, 876–80

Allen, J. P., M. M. Schad, B. Oudekerk and J. Chango (2014), 'What Ever Happened to the "Cool" Kids? Long-Term Sequelae of Early Adolescent Pseudomature Behavior', *Child Development* 85 (5): 1,866–80

Allen, K. L. and R. F. Kay (2012), 'Dietary Quality and Encephalization in Platyrrhine Primates', *Proceedings of the Royal Society of Lonon. Series B, Containing Papers of a Biological Character*, 279 (1,729): 715–21

Alloy, L. B. and L. Y. Abramson (1988), 'Depressive Realism: Four Theoretical Perspectives', in L. B. Alloy (ed.), *Cognitive Processes in Depression* (pp. 223–65), New York, NY: Guilford Press

Amaral, D., D. Geschwind and G. Dawson (eds) (2011), *Autism Spectrum Disorders*, Oxford: Oxford University Press

Amaral, D. G., C. M. Schumann and C. W. Nordahl (2008), 'Neuroanatomy of Autism', *Trends in Neuroscience*, 31: 137–45

Amaral, D. G. (2002), 'The Primate Amygdala and the Neurobiology of Social Behavior: Implications for Understanding Social Anxiety', *Biological Psychiatry*, 51: 11–17

—— (2003), 'The Amygdala, Social Behavior, and Danger Detection', *Annals of the New York Academy of Sciences*, 1,000: 337–47

Amso, D. and G. Scerif (2015), 'The Attentive Brain: Insights from Developmental Cognitive Neuroscience', *Nature Reviews Neuroscience*, 16 (10): 606–19

Andersen, R. A. (1987), 'Inferior Parietal Lobule Function in Spatial Perception and Visuomotor Integration', in F. Plum (ed.), *Handbook of Physiology*, sect. 1, 'The Nervous System', vol. 5, *Higher Functions of the Brain*, pt 2 (pp. 483–518), Bethesda, MD: American Physiological Society

Anderson, A. (2007), 'Feeling Emotional: The Amygdala Links Emotional Perception and Experience', *Social Cognitive and Affective Neuroscience*, 2 (2): 71–2

Anderson, S. W., A. Bechara, H. Damasio et al. (1999), 'Impairment of Social and Moral Behavior Related to Early Damage in Human Prefrontal Cortex', *Nature Neuroscience*, 2: 1,032–7

Anney, R. J. L., L. L. Klei, D. Pinto et al. (2010), 'A Genome-Wide Scan for Common Alleles Affecting Risk for Autism', *Human Molecular Genetics*, 19 (20): 4,072–82

Arac, J. (2002), 'Anglo-Globalism?', *New Left Review*, 16, July/August, https://newleftreview.org/issues/II16/articles/jonathan-arac-anglo-globalism

Aragón, J. L., G. G. Naumis, M. Bai et al. (2008), 'Turbulent Luminance in Impassioned Van Gogh Paintings', *Journal of Mathematical Imaging and Vision*, 30: 275–83

Aristotle, *Politics*, trans. B. Jowett (1885), Mineola, NY: Dover, 2000

Aru, J. and T. Bachmann (2015), 'Still Wanted: The Mechanisms of Consciousness!', *Frontiers in Psychology*, 6, article 5

Aru, J., T. Bachmann, W. Singer and L. Melloni (2012), 'Distilling the Neural Correlates of Consciousness', *Neuroscience and Biobehavioral Reviews*, 36: 737–46

Asperger, H. (1944), '"Autistic Psychopathy" in Childhood', in U. Frith (ed.), *Autism and Asperger Syndrome* (pp. 37–92), Cambridge: Cambridge University Press

Atkinson, A. P., A. S. Heberlein and R. Adolphs (2007), 'Spared Ability to Recognise Fear from Static and Moving Whole-Body Cues Following Bilateral Amygdala Damage', *Neuropsychologia*, 45 (12): 2,772–82

Badre, D. and A. D. Wagner (2007), 'Left Ventrolateral Prefrontal Cortex and the Cognitive Control of Memory', *Neuropsychologia*, 45: 2,883–901

Baer, J., J. C. Kaufman and R. F. Baumeister (2008), *Are We Free? Psychology and Free Will*, New York, NY: Oxford University Press

Bailey, C. H. and M. C. Chen (1983), Morphological Basis of Long-Term Habituation and Sensitization in Aplysia', *Science*, 220: 91–3

Bailey, C. H. and E. R. Kandel (2008), 'Synaptic Remodeling, Synaptic Growth and the Storage of Long-Term Memory in Aplysia', *Progress in Brain Research*, 169: 179–98

Bandes, S. A. and J. A. Blumenthal (2012), 'Emotion and the Law', *Annual Review of Law and Social Science*, 8: 161–81

Barbas, H. (2000), 'Connections Underlying the Synthesis of Cognition, Memory, and Emotion in Primate Prefrontal Cortices', *Brain Research Bulletin*, 52: 319–30

—— (2015), 'General Cortical and Special Prefrontal Connections: Principles from Structure to Function', *Annual Review of Neuroscience*, 38: 269–89

Barbic, S. P., Z. Durisko and P. W. Andrews (2014)., 'Measuring the Bright Side of Being Blue: A New Tool for Assessing Analytical Rumination in Depression', *PLoS ONE*, 9 (11): e112077

Barco, A., C. H. Bailey and E. R. Kandel (2006), 'Common Molecular Mechanisms in Explicit and Implicit Memory', *Journal of Neurochemistry*, 97: 1,520–33

Bargmann, C. I. (2012), 'Beyond the Connectome: How Neuromodulators Shape Neural Circuits', *Bioessays*, 34 (6): 458–65

Baron-Cohen, S. (2006), 'The Hyper-Systemizing, Assortative Mating Theory of Autism', *Progress in Neuro-Psychopharmacology & Biological Psychiatry*, 30 (5): 865–72

—— (2010), 'Empathizing, Systemizing, and the Extreme Male Brain Theory of Autism', *Progress in Brain Research*, 186: 167–75

Baron-Cohen, S., A. Cox, G. Baird et al. (1996), 'Psychological Markers in the Detection of Autism in Infancy in a Large Population', *British Journal of Psychiatry*, 168: 158–63

Baron-Cohen, S., H. Tager-Flusberg and D. J. Cohen (2000), *Understanding Other Minds: Perspectives from Autism*, 2nd edn, Oxford: Oxford University Press

Baron-Cohen, S., A. M. Leslie and U. Frith (1985), 'Does the Autistic Child Have a "Theory of Mind"?' *Cognition*, 21 (1): 37–46

Baron-Cohen, S., S. J. Wheelwright, R. Skinner et al. (2001), 'The Autism -Spectrum Quotient (AQ): Evidence from Asperger Syndrome/High-Functioning Autism, Males and Females, Scientists and Mathematicians', *Journal of Autism and Developmental Disorders*, 31 (1): 5–17

Barrett, L. F. and B. S. Ajay (2013), 'Large-Scale Brain Networks in Affective and Social Neuroscience: Towards an Integrative Functional Architecture of the Brain', *Current Opinion in Neurobiology*, 23 (3): 361–72

Barrett, L. F. and W. K. Simmons (2015), 'Interoceptive Predictions in the Brain', *Nature Reviews Neuroscience*, 16 (7): 419–29

Barton, R. A. and C. Venditti (2013), 'Human Frontal Lobes Are Not Relatively Large', *Proceedings of the National Academy of Sciences of the United States of America*, 110 (22): 9,001–6

Basser, P. J., J. Mattiello and D. LeBihan (1994), 'MR Diffusion Tensor Spectroscopy and Imaging', *Biophysical Journal*, 66: 259–67

Bates, E., B. Wulfeck and B. MacWhinney (1991), 'Cross-Linguistic Research in Aphasia: An Overview', *Brain and Language*, 41: 123–48

Batista, A. P., C. A. Buneo, L. H. Snyder and R. A. Andersen (1999), 'Reach Plans in Eye-Centered Coordinates', *Science* 285: 257–60.

Bauby, J. (2008), *The Diving-Bell and the Butterfly*, HarperCollins

Bauman, M. L. (1999), 'Autism: Clinical Features and Neurobiological Observations', in: H. Tager-Flusberg (ed), *Neurodevelopmental Disorders* (Developmental Cognitive Neuroscience), pp. 383–99, Cambridge, MA: MIT Press

Baumeister, R. F., E. J. Masicampo and C. N. DeWall (2009), 'Prosocial Benefits of Feeling Free: Disbelief in Free Will Increases Aggression and Reduces Helpfulness', *Personality and Social Psychology Bulletin*, 35, 260–8

Bayne, T., A. Cleeremans and P. Wilken (2009), *The Oxford Companion to Consciousness*, Oxford: Oxford University Press

Baynes, K. (1990), 'Language and Reading in the Right Hemisphere: Highways or Byways of the Brain?' *Journal of Cognitive Neuroscience*, 2: 159–79

Beach, S. R. H., G. H. Brody, T. D. Gunter et al. (2010), 'Child Maltreatment Moderates the Association of MAOA with Symptoms of Depression and Antisocial Personality Disorder', *Journal of Family Psychology*, 24 (1): 12–20

Bear, M. F., K. M. Huber and S. T. Warren (2004), 'The mGluR Theory of Fragile X Syndrome', *Trends in Neuroscience*, 27: 370–7

Beard, M. (2014), *Laughter in Ancient Rome: On Joking, Tickling, and Cracking Up* (Sather Classical Lectures), Berkeley, CA: University of California Press

Beaton, M. (2005). 'What RoboDennett Still Doesn't Know', *Journal of Consciousness Studies*, 12: 3–25

Bechara, A., A. R. Damasio, H. Damasio and S. W. Anderson (1994), 'Insensitivity to Future Consequences Following Damage to Human Prefrontal Cortex', *Cognition*, 50: 7–15

Bechara, A., H. Damasio and A. R. Damasio (2000), 'Emotion, Decision-Making and the Orbitofrontal Cortex', *Cerebral Cortex*, 10: 295–307

Bechara, A., H. Damasio, D. Tranel and A. R. Damasio (1997), 'Deciding Advantageously Before Knowing the Advantageous Strategy', *Science*, 275: 1,293–5

Bechara, A., D. Tranel, H. Damasio et al. (1995), 'A Double Dissociation of Conditioning and Declarative Knowledge Relative to the Amygdala and Hippocampus in Humans', *Science*, 269: 1,115–18

Beckers, G. J. L., J. J. Bolhuis, K. Okanoya and R. C. Berwick (2012), 'Birdsong Neurolinguistics: Songbird Context-Free Grammar Claim is Premature', *Neuroreport*, 23 (3): 139–45

Bemis, D. K. and L. Pylkkänen (2011), 'Simple Composition: A Magnetoencephalography Investigation into the Comprehension of Minimal Linguistic Phrases', *Journal of Neuroscience*, 31 (8): 2,801–14

Ben-Shachar, M., R. F. Dougherty and B. A. Wandall (2007), 'White Matter Pathways in Reading', *Current Opinion in Neurobiology*, 17: 258–70

Ben-Shachar, M., D. Palti and Y. Grodzinsky (2004), 'Neural Correlates of Syntactic Movement: Converging Evidence from Two fMRI Experiments', *NeuroImage*, 21 (4): 1,320–36

Benito-Kwiecinski, S., S. Giandomenico, M. Sutcliffe et al. (2021), 'An Early Cell Shape Transition Drives Evolutionary Expansion of the Human Forebrain', *Cell*, 184 (8): 2,084–102

Bertossa, F., M. Besa, R. Ferrari and F. Ferri (2008), 'Point Zero: A Phenomenological Inquiry into the Seat of Consciousness', *Perceptual and Motor Skills*, 107 (2): 323–35

Best, C., S. Arora, F. Porter and M. Doherty (2015), 'The Relationship Between Subthreshold Autistic Traits, Ambiguous Figure Perception and Divergent Thinking', *Journal of Autism and Developmental Disorders*, 45: 4,064–73

Binder, J. R., R. H. Desai, W. W. Graves and L. L. Conant (2009), 'Where is the Semantic System? A Critical Review and Meta-Analysis of 120 Functional Neuroimaging Studies', *Cerebral Cortex*, 19 (12): 2,767–96

Binet, A. (1916), 'New Methods for the Diagnosis of the Intellectual Level of Subnormals', *The Development of Intelligence, in Children: The Binet–Simon Scale* (pp. 37–90) trans. E. S. Kite, Baltimore, MD: Williams & Wilkins

Bishop, D. V. M. (1983), 'Linguistic Impairment After Left Hemi-Decortication for Infantile Hemiplegia: A Reappraisal', *Quarterly Journal of Experimental Psychology*, 35A: 199–207

Bishop, N. A., T. Lu and B. A. Yankner (2010), 'Neural Mechanisms of Ageing and Cognitive Decline', *Nature*, 464 (7,288): 529–35

Bisiach, E. and C. Luzzatti (1978), 'Unilateral Neglect of Representational Space', *Cortex*, 14: 129–33

Bisley, J. W. and M. E. Goldberg (2010), 'Attention, Intention, and Priority in the Parietal Lobe', *Annual Review of Neuroscience*, 33: 1–21

Blackmore, S. (2005), *Conversations on Consciousness*, New York, NY: Oxford University Press

—— (2010), 'Memetics Does Not Provide a Useful Way of Understanding Cultural Evolution', in F. Ayala and R. Arp (eds), *Contemporary Debates in Philosophy of Biology* (pp. 255–72, 273–91), Chichester: Wiley–Blackwell

—— (2011), *Zen and the Art of Consciousness*, Oneworld

—— (2017), 'Untestable Claims and the Evolution of Consciousness', *Trends in Ecology and Evolution*, 32 (5): 311–12

Blackmore, S. J. (2002), 'There is No Stream of Consciousness', *Journal of Consciousness Studies*, 9 (5–6): 17–28

Blakemore, S. and E. Troscianoko (2018), *Consciousness: An Introduction*, Routledge

Block, N. (2005), 'Two Neural Correlates of Consciousness', *Trends in Cognitive Sciences*, 9 (2): 46–52

Bloom, F. and A. Lazerson (1988), *Brain, Mind and Behavior*, 2nd edn, New York, NY: Freeman

Boback, S. M., C. L. Cox, B. D. Ott et al. (2007), 'Cooking and Grinding Reduces the Cost of Meat Digestion', *Comparative Biochemistry and Physiology, A: Molecular and Integrative Physiology*, 148 (3): 651–6

Boks, P. (2003), 'To be Two or Not to Be', in P. Broks, *Into the Silent Land: Travels in Neuropsychology* (pp. 204–25), Atlantic

Boron, Z. (2014), 'How Powerful is a Mind? Supercomputer Takes 40 Minutes to Map 1 Second of Brain Activity', *Independent*, 14 January 2014, www.independent.co.uk/life-style/gadgets-and-tech/how-powerful-is-a-mind-supercomputer-takes-40-minutes-to-map-1-second-of-brain-activity-9059225.html

Botha, R. and C. Knight (eds) (2009), *The Prehistory of Language*, New York, NY: Oxford University Press

Bourgeron, T. (2015), 'From the Genetic Architecture to Synaptic Plasticity in Autism Spectrum Disorder', *Nature Reviews Neuroscience*, 16 (9): 551–63

Boxer Wachler, B. S. (2017), *Perceptual Intelligence: The Brain's Secret to Seeing Past Illusion, Misperception, and Self-Deception*, Novato, CA: New World Library

Boynton, G. M., S. A. Engel, G. H. Glover and D. J. Heeger (1996), 'Linear Systems Analysis of Functional Magnetic Resonance Imaging in Human V1', *Journal of Neuroscience*, 16: 4,207–21

Brashears, M. E. (2011), 'Small Networks and High Isolation? A Reexamination of American Discussion Networks', *Social Networks*, 33 (4): 331–41

Breiter, H. C., I. Aharon, D. Kahneman et al. (2001), 'Functional Imaging of Neural Responses to Expectancy and Experience of Monetary Gains and Losses', *Neuron* 30: 619–39

Brewer, J. A., P. D. Worhunksy, J. R. Gray et al. (2011) 'Meditation Experience is Associated with Differences in Default Mode Network Activity and Connectivity', *Proceedings of the National Academy of Sciences of the United States of America*, 108 (50): 20,254–9

Broca, P. (1861), 'Remarques sur le siège de la faculté du langage articulé, suivies d'une observation d'aphémie (perte de la parole)', *Bull Société Anatomique de Paris*, 6: 330–57

—— (1878), 'Anatomie comparée des circonvolutions cérébrales: Le grand lobe limbique et le scissure limbique dans le série des mammitères', *Annual Review of Anthropology*, 12: 646–57

Brodmann, K. (1909), *Vergleichende Lokalisationslehre der Grosshirnrinde in ihren Prinzipien dargestellt auf Grund des Zellenbaues*, Leipzig: Barth

Brown, K. (2016), An Interview with Jürgen Knoblich, the Company of Biologists, www.youtube.com/watch?v=hhmh8YKDwyo

Brown, P., T. Sutikna, M. J. Morwood et al. (2004), 'A New Small-Bodied Hominin from the Late Pleistocene of Flores, Indonesia', *Nature*, 431 (7,012): 105,561

Brunet, M., F. Guy, D. Pilbeam et al. (2002), 'A New Hominid from the Upper Miocene of Chad, Central Africa', *Nature*, 418 (6,894): 145–51

Buchsbaum, B. R., J. Baldo, K. Okada et al. (2011), 'Conduction Aphasia, Sensory-Motor Integration, and Phonological Short-Term Memory: An Aggregate Analysis of Lesion and fMRI Data', *Brain and Language*, 119: 119–28

Buckner, R. L., P. A. Bandettini, K. M. O'Craven et al. (1996). 'Detection of Cortical Activation During Averaged Single Trials of a Cognitive Task Using Functional Magnetic Resonance Imaging', *Proceedings of the National Academy of Sciences of the United States of America*, 93: 14,878–883

Buckner, R. L., A. Z. Snyder, M. E. Raichle and J. C. Morris (2000), 'Functional Brain Imaging of Young, Nondemented, and Demented Older Adults', *Journal of Cognitive Neuroscience*, 12, Suppl. 2: 24–34

Bullmore, E. (2019), *The Inflamed Mind: A Radical New Approach to Depression*, Short Books

Bunn, H. (2012), talk given at the European Society for the Study of Human Evolution

Burt, R. S. (2004), 'Structural Holes and Good Ideas', *American Journal of Sociology*, 110 (2): 349–99

Bushnell, M. C., M. E. Goldberg and D. L. Robinson (1981), 'Behavioral Enhancement of Visual Responses in Monkey Cerebral Cortex, I, Modulation in Posterior Parietal Cortex Related to Selective Visual Attention', *Journal of Neurophysiology*, 46: 755–72

Buxton, R. B., E. C. Wong and L. R. Frank (1998), 'Dynamics of Blood Flow and Oxygenation Changes During Brain Activation: The Balloon Model', *Magnetic Resonance in Medicine*, 39: 855–64

Cahill, L. and J. L. McGaugh (1998), 'Mechanisms of Emotional Arousal and Lasting Declarative Memory', *Trends in Neuroscience*, 21: 294–9

Cairó, O. (2011), 'External Measures of Cognition', *Frontiers in Human Neuroscience*, 5, article 108

Capek, C. M., G. Grossi, A. J. Newman et al. (2009), 'Brain Systems Mediating Semantic and Syntactic Processing in Deaf Native Signers: Biological Invariance and Modality Specificity', *Proceedings of the National Academy of Sciences of the United States of America*, 106 (21): 8,784–9

Carruthers, P. (2004), 'Suffering Without Subjectivity', *Philosophical Studies*, 121: 99–125

Casey, N. (2002), *Unholy Ghost: Writers on Depression*, New York, NY: William Morrow

Cassidy, S. B., and C. A. Morris (2002), 'Behavioral Phenotypes in Genetic Syndromes: Genetic Clues to Human Behavior', in L. A. Barness (ed.), *Advances in Pediatrics*, vol. 49 (pp. 59–86), Philadelphia, PA: Mosby

Castelli, F., F. Happé, C. D. Frith and U. Frith (2002), 'Autism, Asperger Syndrome and Brain Mechanisms for the Attribution of Mental States to Animated Shapes', *Brain*, 125: 1,839–49

Catani, M. and M. Thiebaut de Schotten (2008), 'A Diffusion Tensor Imaging Tractography Atlas for Virtual in Vivo Dissections', *Cortex*, 44 (8): 1,105–32

Cave, S. (2016), 'There's No Such Thing as Free Will – But We're Better Off Believing in It Anyway', *Atlantic*, June, www.theatlantic.com/magazine/archive/2016/06/theres-no-such-thing-as-free-will/480750/

Cervera, M. A., S. R. Soekadar, J. Ushibaet et al. (2018), 'Brain–Computer Interfaces for Post-Stroke Motor Rehabilitation: A Meta-Analysis', *Annals of Clinical and Translational Neurology*, 5 (5): 651–63

Chalmers, D. J. (1995b), 'The Puzzle of Conscious Experience', *Scientific American*, December, 62–8

Chanes, L. and L. F. Barrett (2016), 'Redefining the Role of Limbic Areas in Cortical Processing', *Trends in Cognitive Sciences*, 20 (2): 96–106

Chang, L., S. Zhang, M. M. Poo and N. Gong (2017), 'Spontaneous Expression of Mirror Self-Recognition in Monkeys After Learning Precise Visual-Proprioceptive Association for Mirror Images', *Proceedings of the National Academy of Sciences of the United States of America*, 114 (12): 3,258–63

Charter, N. (2018), *The Mind is Flat: The Illusion of Mental Depth and the Improvised Mind*, Allen Lane

Chelly, J., J.-L. Mandel (2001), 'Monogenic Causes of X-Linked Mental Retardation', *Nature Reviews Genetics*, 2: 669–80

Cheney, D. L. and R. M. Seyfarth (1990), *How Monkeys See the World: Inside the Mind of Another Species*, Chicago, IL: University of Chicago Press

Cheng, K., R. A. Waggoner and K. Tanaka (2001), 'Human Ocular Dominance Columns as Revealed by High-Field Functional Magnetic Resonance Imaging', *Neuron*, 32: 359–74

Choi, G. B., H. W. Dong, A. J. Hurphy et al. (2005), 'Lhx6 Delineates a Pathway Mediating Innate Reproductive Behaviors from the Amygdala to the Hypothalamus', *Neuron* 19: 647–60

Chomsky N. (1959), 'A Review of B. F. Skinner 's "Verbal Behavior"', *Language*, 35: 26–58

—— (1968), 'Language and the Mind', *Psychology Today*, 1: 48–68

Christopoulos, G. I., P. N. Tobler, P. Bossaerts et al. (2009), 'Neural Correlates of Value, Risk, and Risk Aversion Contributing to Decision Making Under Risk', *Journal of Neuroscience*, 29: 12,574–83

Churchland, P. M. (1985), 'Reduction, Qualia, and the Direct Introspection of Brain States', *Journal of Philosophy*, 82 (1): 8–28

Churchland, P. S. (1996), 'The Hornswoggle Problem', *Journal of Consciousness Studies*, 3: 402–8

Clare, R., V. G. King, M. Wirenfeldt and H. V. Vinters (2010), 'Synapse Loss in Dementias', *Journal of Neuroscience Research*, 88 (10): 2,083–90

Claridge-Chang, A., R. D. Roorda, E. Vrontou et al. (2009), 'Writing Memories with Light-Addressable Reinforcement Circuitry', *Cell*, 139: 405–15

Clark, A. (2015), *Surfing Uncertainty: Prediction, Action, and the Embodied Mind*, New York, NY: Oxford University Press

Clynes, M. E. and N. S. Kline (1960), 'Cyborgs and Space', *Astronautics*, September 1960: 26–7, 74–6

Cochran, D. M., D. Fallon, M. Hill and J. A. Frazier (2013), 'The Role of Oxytocin in Psychiatric Disorders: A Review of Biological and Therapeutic Research Findings', *Harvard Review of Psychiatry*, 21 (5): 219–47

Cohen, L., A. Jobert, D. L. Bihan and S. Dehaene (2004), 'Distinct Unimodal and Multimodal Regions for Word Processing in the Left Temporal Cortex', *NeuroImage*, 23 (4): 1,256–70

Colby, C. L. J. R. Duhamel and M. E. Goldberg (1996), 'Visual, Presaccadic and Cognitive Activation of Single Neurons in Monkey Lateral Intraparietal Area', *Journal of Neurophysiology*, 76: 2,841–52

Corbetta, M., F. M. Miezin, G. L. Shulman and S. E. Petersen (1993), 'A PET Study of Visuospatial Attention', *Journal of Neuroscience*, 13: 1,202–26

Corbetta, M. and G. L. Shulman (2002), 'Control of Goal-Directed and Stimulus-Driven Attention in the Brain', *Nature Reviews Neuroscience*, 3: 201–15

Corkin, S. (2002), 'What's New with the Amnesic Patient H.M.?', *Nature Reviews Neuroscience*, 3: 153–60

Corkin, S., D. G. Amaral, R. G. González et al. (1997), 'H.M.'s Medial Temporal Lobe Lesion: Findings from Magnetic Resonance Imaging', *Journal of Neuroscience*, 17: 3,964–79

Cornell, T. L., V. A. Fromkin and G. Mauner (1993), 'A Linguistic Approach to Language Processing in Broca's Aphasia: A Paradox Resolved', *Current Directions in Psychological Science*, 2: 47–52

Courchesne, E., K. Pierce, C. M. Schumann et al. (2007), 'Mapping Early Brain Development in Autism', *Neuron*, 56: 399–413

Craig, A. D. (2002), 'How Do You Feel? Interoception: The Sense of the Physiological Condition of the Body', *Nature Reviews Neuroscience*, 3: 655–66

—— (2009), 'How Do You Feel – Now? The Anterior Insula and Human Awareness', *Nature Reviews Neuroscience*, 10: 59–70

—— (2015), *How Do You Feel? An Interoceptive Moment with Your Neurobiological Self*, Princeton, NJ: Princeton University Press

Crick, F. (1995), *The Astonishing Hypothesis: The Scientific Search for the Soul*, New York, NY: Touchstone/Simon & Schuster

Crick, F. and C. Koch (1990), 'Towards a Neurobiological Theory of Consciousness', *Seminars in Neuroscience*, 2: 263–75

—— (1995). 'Are We Aware of Neural Activity in Primary Visual Cortex?', *Nature*, 375: 121–3

—— (2003), 'A Framework for Consciousness', *Nature Neuroscience*, 6: 119–26

Critchley, H. D. (2005), Neural Mechanisms of Autonomic, Affective, and Cognitive Integration', *Journal of Comparative Neurology*, 493: 154–66

Critchley, M. (1953), *The Parietal Lobes*, New York, NY: Hafner

Crivelli, C., J. A. Russell, S. Jarillo and J. M. Fernandez-Dols (2016), 'The Fear Gasping Face as a Threat Display in a Melanesian Society', *Proceedings of the National Academy of Sciences of the United States of America*, 113 (44): 12,403–7

Curtiss, S. (2016), interview with Rory Carroll in the *Guardian*: 'Starved, Tortured, Forgotten: Genie, The Feral Child Who Left a Mark on Researchers, 14 July 2016, www.theguardian.com/society/2016/jul/14/genie-feral-child-los-angeles-researchers

Dahlström, A. and A. Carlsson (1986), 'Making Visible the Invisible', in M. J. Parnam and J. Bruinnvels (eds), *Discoveries in Pharmacology*, vol. 3, *Pharmacological Methods, Receptors and Chemotherapy* (pp. 97–125), Amsterdam: Elsevier

Damasio, A. R. (1992), 'Aphasia', *New England Journal of Medicine*, 326: 531–9

—— (1996), 'The Somatic Marker Hypothesis and the Possible Functions of the Prefrontal Cortex', *Philosophical Transactions of the Royal Society of London: B. Biological Sciences*, 351: 1,413–20

—— (2006), *Descartes' Error: Emotion, Reason, and the Human Brain*, Vintage

—— (2010), *Self Comes to Mind*, New York, NY: Pantheon

Damasio, A. R., T. J. Grabowski, A. Becharaet et al. (2000), 'Subcortical and Cortical Brain Activity During the Feeling of Self-Generated Emotions', *Nature Neuroscience*, 3: 1,049–56

Damasio, A. R. and D. Tranel (1993), 'Nouns and Verbs Are Retrieved with Differently Distributed Neural Systems', *Proceedings of the National Academy of Sciences of the United States of America*, 90: 4,957–60

Damasio, H., T. Grabowski, R. Frank et al. (1994), 'The Return of Phineas Gage: Clues About the Brain from the Skull of a Famous Patient', *Science*, 264: 1,102–5

Damasio, H., T. J. Grabowski, D. Tranel et al. (1996), 'A Neural Basis for Lexical Retrieval', *Nature*, 380: 499–505

Damasio, H., D. Tranel, T. J. Grabowski et al. (2004), 'Neural Systems Behind Word and Concept Retrieval', *Cognition*, 92: 179–229

Darian-Smith, I. (1982), 'Touch in Primates', *Annual Review of Psychology*, 33: 155–94

Darwin, C. (1871), *The Descent of Man, and Selection in Relation to Sex*, John Murray

—— (1872), *Emotions in Man and Animals*, John Murray

—— (1872), *The Expression of the Emotions in Man and Animals*, Stilwell, KS: Digireads.com

Davis, M. and P. J. Whalen (2001), 'The Amygdala: Vigilance and Emotion', *Molecular Psychiatry*, 6: 13–34

Dawson, G., A. N. Meltzoff, J. Osterling et al. (1998), 'Children with Autism Fail to Orient to Naturally Occurring Social Stimuli', *Journal of Autism and Developmental Disorders*, 28 (6): 479–85

Dawson, G., K. Toth, R. Abbott et al. (2004), 'Early Social Attention Impairments in Autism: Social Orienting, Joint Attention, and Attention to Distress', *Developmental Psychology*, 40: 271–83

Day, J. J. and J. D. Sweatt (2011), 'Epigenetic Mechanisms in Cognition', *Neuron*, 70: 813–29

De Grey, A. (2008), *Ending Aging: The Rejuvenation Breakthroughs That Could Reverse Human Aging in Our Lifetime*, New York, NJ: St Martin's Griffin

De Waal, F. (2017), *Are We Smart Enough to Know How Smart Animals Are?* Granta

Decety, J. (2010), 'The Neurodevelopment of Empathy in Humans', *Developmental Neuroscience*, 32 (4), 257–67

Decety, J., I. B.-A. Bartal, F. Uzefovsky and A. Knafo-Noam (2015), 'Empathy as a Driver of Prosocial Behaviour: Highly Conserved Neurobehavioural Mechanisms Across Species', *Philosophical Transactions of the Royal Society of London: B. Biological Sciences*, 371 (1,686): 20,150,077

Dehaene, S. and J.-P. Changeux (2011), 'Experimental and Theoretical Approaches to Conscious Processing', *Neuron*, 70: 201–27

Dehaene, S., H. Lau and S. Kouider (2017), 'What is Consciousness, and Could Machines Have It?', *Science*, 358 (6,362): 486–92

Dehaene, S. and L. Naccache (2001), 'Towards a Cognitive Neuroscience of Consciousness: Basic Evidence and a Workspace Framework', *Cognition*, 79: 1–37.

Dennett, D. (1991), *Consciousness Explained*, Boston, MA: Little, Brown

Dennett, D. C. (2017), *From Bacteria to Bach and Back: The Evolution of Minds*, Allen Lane

Dennis, M. and H. A. Whitaker (1976), 'Language Acquisition Following Hemidecortication: Linguistic Superiority of the Left Over the Right Hemisphere', *Brain and Language*, 3: 404–33

Devlin, H. (2018), 'Thousands of Autistic Girls and Women "Going Undiagnosed" Due to Gender Bias', *Guardian*, 14 September 2018, www.theguardian.com/society/2018/sep/14/thousands-of-autistic-girls-and-women-going-undiagnosed-due-to-gender-bias

DiLuca, M. and J. Olesen (2014), 'The Cost of Brain Diseases: A Burden or a Challenge?', *Neuron*, 82 (6): 1,205–8

Doi, T. (1971, 2002), *The Anatomy of Dependence*, New York, NY: Kodansha America

Dolan, R. J. (2002), 'Emotion, Cognition, and Behavior', *Science*, 298: 1,191–4

Donahue, M. Z. (2017), 'How a Color-Blind Artist Became the World's First Cyborg', *National Geographic*, www.nationalgeographic.com/news/2017/04/worlds-first-cyborg-human-evolution-science/

Doupe, A. and P. K. Kuhl (1999), 'Birdsong and Speech: Common Themes and Mechanisms', *Annual Reviews of Neuroscience*, 22: 567–631

Draaisma, D. and P. Vincent (2000), *Metaphors of Memory: A History of Ideas About the Mind*, Cambridge: Cambridge University Press

Drachman, D. A. and J. Leavitt (1974), 'Human Memory and the Cholinergic System: A Relationship to Aging?', *Archives of Neurology*, 30 (2): 113–21

Drevets, W. C., J. L. Price, J. R. Simpson Jr, et al. (1997), 'Subgenual Prefrontal Cortex Abnormalities in Mood Disorders', *Nature*, 386: 769–70

Dronkers, N. F. (1996), 'A New Brain Region for Coordinating Speech Articulation', *Nature*, 384: 159–61

Duffau, H., P. Gatignol, S. Moritz-Gasserand, E. Mandonnet (2009), 'Is the Left Uncinate Fasciculus Essential for Language?', *Journal of Neurology*, 256 (3): 382–9

Duhamel, J. R., C. L. Colby and M. E. Goldberg (1998), 'Ventral Intraparietal Area of the Macaque: Congruent Visual and Somatic Response Properties', *Journal of Neurophysiology*, 79: 126–36

Dunbar, R. I. M. (1992), 'Neocortex Size as a Constraint on Group Size in Primates', *Journal of Human Evolution*, 22 (6): 469–93

Durand, C. M., C. Betancur, T. M. Boeckers et al. (2007), 'Mutations in the Gene Encoding the Synaptic Scaffolding Protein SHANK3 are Associated with Autism Spectrum Disorders', *Nature Genetics*, 39 (1): 25–7

Durdiaková, J., V. Warrier, S. Banerjee-Basuet et al. (2014), 'STX1A and Asperger Syndrome: A Replication Study', *Molecular Autism*, 5 (1): 14

Durdiaková, J., V. Warrier, S. Baron-Cohen and B. Chakrabarti (2014), 'Single Nucleotide Polymorphism rs6716901 in SLC25A12 Gene is Associated with Asperger Syndrome', *Molecular Autism*, 5 (1): 25

Easton Ellis, B. (2011), *American Psycho*, Picador

Edelman, D. B. and A. K. Seth (2009), 'Animal Consciousness: A Synthetic Approach', *Trends in Neurosciences*, 32 (9), 476–84

Edelman, D. B., B. J. Baars and A. K. Seth (2005), 'Identifying Hallmarks of Consciousness in Non-Mammalian Species', *Consciousness and Cognition*, 14 (1): 169–87

Edelson, M., T. Sharot, R. J. Dolan and Y. Dudai (2011), 'Following the Crowd: Brain Substrates of Long-Term Memory Conformity', *Science*, 333 (6,038): 108–111.

Egorov, A. V., B. N. Hamam, E. Fransen et al. (2002), 'Graded Persistent Activity in Entorhinal Cortex Neurons', *Nature*, 420: 173–8

Eimas, P. D., E. R. Siqueland, P. Jusczyk and J. Vigorito (1971), 'Speech Perception in Infants', *Science*, 171: 303–6

Ekman, P. (2004), *Emotions Revealed: Understanding Faces and Feelings*, Weidenfeld & Nicolson

Eldridge, L. L., B. J. Knowlton, C. S. Furmanskiet et al. (2000), 'Remembering Episodes: A Selective Role for the Hippocampus During Retrieval', *Nature Neuroscience*, 3: 1,149–52

Eliot, G. (1871), *Middlemarch*, Penguin Classics, 2011

Ellis, S. E., R. Panitch, A. B. West and D. E. Arking (2016), 'Transcriptome Analysis of Cortical Tissue Reveals Shared Sets of Downregulated Genes in Autism and Schizophrenia', *Translational Psychiatry*, 6 (5): e817

Engel, A. K. (2003), Temporal Binding and the Neural Correlates of Consciousness', in A. Cleeremans (ed.), The *Unity of Consciousness: Binding, Integration and Dissociation* (pp. 132–52), New York, NY: Oxford University Press

Eslinger, P. J. and A. R. Damasio (1985), 'Severe Disturbance of Higher Cognition After Bilateral Frontal Lobe Ablation: Patient EVR', *Neurology*, 35: 1,731–41

Everett, D. (2018), interview with Nicola Davis for the *Guardian*, 'Homo erectus May Have Been a Sailor – and Able to Speak', 20 February 2018, www.theguardian.com/science/2018/feb/20/homo-erectus-may-have-been-a-sailor-and-able-to-speak

Everitt, B. J., R. N. Cardinal, J. A. Parkinson and T. W. Robbins (2003), 'Appetitive Behavior: Impact of Amygdala-Dependent Mechanisms of Emotional Learning', *Annals of the New York Academy of Sciences*, 985: 233–5

Falck, B., N. A. Hillarp, G. Thieme and A. Torp (1962), 'Fluorescence of Catecholamines and Related Compounds Condensed with Formaldehyde', *Journal of Histochemistry and Cytochemistry*, 10: 348–54

Falk, D. (2009), 'The Natural Endocast of Taung (Australopithecus

Africanus): Insights from the Unpublished Papers of Raymond Arthur Dart', *American Journal of Physical Anthropology*, 140, Suppl. 49: 49–65

Falkner, A. L., L. Grosenick, T. J. Davidson et al. (2016), 'Hypothalamic Control of Male Aggression-Seeking Behavior', *Nature Neuroscience*, 19 (4): 596–604

Fanselow, M. S. and A. M. Poulos (2005), 'The Neuroscience of Mammalian Associative Learning', *Annual Review of Psychology*, 56: 207–34

Feinberg, T. E. (2009), *From Axons to Identity: Neurological Explorations of the Nature of Self*, New York, NY: W. W. Norton

Feinstein, J. S., R. Adolphs, A. R. Damasio and D. Tranel (2011), 'The Human Amygdala and the Induction and Experience of Fear', *Current Biology*, 21 (1): 34–8

Feldman Barrett, L. (2017), *How Emotions Are Made: The Secret Life of the Brain*, Macmillan

Felleman, D. J. and D. C. Van Essen (1991), 'Distributed Hierarchical Processing in the Primate Cerebral Cortex', *Cerebral Cortex*, 1: 1–47

Feltz, A. and F. Cova (2014), 'Moral Responsibility and Free Will: A Meta -Analysis', *Consciousness and Cognition*, 30: 234–46

Fenn, V. (2018), 'These 10 Artists Prove Autism is No Barrier to Creativity', *Metro*, 21 April 2018, https://metro.co.uk/2018/04/21/these-10-artists-prove-autism-is-no-barrier-to-creativity-7446895/

Fernald, A. and P. Kuhl (1987), 'Acoustic Determinants of Infant Preference for Motherese Speech', *Infant Behavior and Development*, 10: 279–93

Fernald, A., V. A. Marchman and A. Weisleder (2013), 'SES Differences in Language Processing Skill and Vocabulary Are Evident at 18 Months', *Developmental Science*, 16 (2): 234–48

Feuillet, L., H. Dufour and J. Pelletier (2007), 'Brain of a White-Collar Worker', *Lancet*, 370 (9,583): 262

Fields, H. L. and E. B. Margolis (2015), 'Understanding Opioid Reward', *Trends in Neurosciences*, 38 (4): 217–25

Fields, R. D. (2011), *The Other Brain: The Scientific and Medical Breakthroughs That Will Heal Our Brains and Revolutionize Our Health*, New York, NY: Simon & Schuster

Filevich, E., P. Vanneste, M. Brass et al. (2013), 'Brain Correlates of Subjective Freedom of Choice', *Consciousness and Cognition*, 22: 1,271–84

Finger, S. (2001), *Origins of Neuroscience: A History of Explorations into Brain Function*, New York, NY: Oxford University Press

Fink, G. R., P. W. Halligan, J. C. Marshall et al. (1996), 'Where in the Brain

Does Visual Attention Select the Forest and the Trees?', *Nature*, 382: 626–8

Fischer, A., F. Sananbenesi, X. Wang et al. (2007), 'Recovery of Learning and Memory is Associated with Chromatin Remodelling', *Nature*, 447: 178–82

Fishwick, S. (2016), 'Spooky London: How Britain's Spy Agencies Are Using New and Unexpected Methods to Recruit the Next Generation', *Evening Standard*, 25 February 2016, www.standard.co.uk/lifestyle/esmagazine/spooky-london-the-new-methods-of-britains-spies-a3189396.html

Fitch, W. T. (2010), *The Evolution of Language*, Cambridge: Cambridge University Press

—— (2016), interview with Carl Zimmer in the *New York Times*, 'Monkeys Could Talk, but They Don't Have the Brains for It', www.nytimes.com/2016/12/09/science/monkeys-speech.html

Fitch, W. T., B. de Boer, N. Mathur and A. A. Ghazanfar (2016), 'Monkey Vocal Tracts Are Speech-Ready', *Science Advances*, 2 (12): e1600723

Flege, J. E. (1995), 'Second Language Speech Learning: Theory, Findings, and Problems', in W. Strange (ed.), *Speech Perception and Linguistic Experience* (pp. 233–77), Timonium, MD: York Press

Flege, J. E., G. H. Yeni-Komshian and S. Liu (1999), 'Age Constraints on Second-Language Acquisition', *Journal of Memory and Language*, 41: 78–104

Florio, M., M. Albert, E. Taverna et al. (2015), 'Human-Specific Gene ARHGAP11B Promotes Basal Progenitor Amplification and Neocortex Expansion', *Science*, 347 (6,229): 1,465–70

Flynn, J. (2013), 'The Flynn Effect and Flynn's Paradox', *Intelligence*, 41: 851–7

Folstein, S. E. and B. Rosen-Sheidley (2001), 'Genetics of Autism: Complex Etiology for a Heterogeneous Disorder', *Nature Reviews Genetics*, 2: 943–55

Fombonne, E. (2009), 'Epidemiology of Pervasive Developmental Disorders', *Pediatric Research*, 65: 591–8

Ford, B. Q. and M. Tamir (2012), 'When Getting Angry is Smart: Emotional Preferences and Emotional Intelligence', *Emotion*, 12 (4): 685–9

Forgas, J. (2007), 'When Sad is Better Than Happy: Negative Affect Can Improve the Quality and Effectiveness of Persuasive Messages and Social Influence Strategies', *Journal of Experimental Social Psychology*, 43 (4): 513–28

Forgas, J., L. Goldenberg and C. Unkelbach (2009), 'Can Bad Weather Improve Your Memory? An Unobtrusive Field Study of Natural Mood Effects on Real-Life Memory', *Journal of Experimental Social Psychology*, 45 (1): 254–7

Foster, J. K. (2008), *Memory: A Very Short Introduction*, Oxford: Oxford University Press

Fox, P. T. and M. E. Raichle (1986), 'Focal Physiological Uncoupling of Cerebral Blood Flow and Oxidative Metabolism During Somatosensory Stimulation in Human Subjects', *Proceedings of the National Academy of Sciences of the United States of America*, 83: 1,140–4

Fox, P. T., M. E. Raichle, M. A. Mintun and C. Dence (1988), 'Nonoxidative Glucose Consumption During Focal Physiologic Neural Activity', *Science*, 241: 462–4

Frans, E. M., S. Sandin, A. Reichenberg et al. (2013), 'Autism Risk Across Generations: A Population-Based Study of Advancing Grandpaternal and Paternal Age', *JAMA Psychiatry*, 70 (5), 516–21

Freud, S. (1915), *The Unconscious* [Standard Edition 14: 159–204], Hogarth Press

—— (1923), *The Ego and the Id* [Standard Edition 19: 1–59], Hogarth Press

Freund, T. F. and G. Buzsaki (1996), 'Interneurons of the Hippocampus', *Hippocampus*, 6: 347–470

Friederici, A. D. (2002), 'Towards a Neural Basis of Auditory Sentence Processing', *Trends in Cognitive Sciences*, 6 (2): 78–84

—— (2005), 'Neurophysiological Markers of Early Language Acquisition: From Syllables to Sentences', *Trends in Cognitive Sciences*, 9 (10): 481–8

—— (2009a), 'Pathways to Language: Fiber Tracts in the Human Brain', *Trends in Cognitive Sciences*, 13 (4): 175–81

—— (2009b), 'Allocating Function to Fiber Tracts: Facing Its Indirectness', *Trends in Cognitive Sciences*, 13 (9): 370–1

—— (2011), 'The Brain Basis of Language Processing: From Structure to Function', *Physiological Reviews*, 91 (4): 1,357–92

—— (2017), *Language in Our Brain: The Origins of a Uniquely Human Capacity*, Cambridge, MA: MIT Press

Friederici, A. D., M. Makuuchi and J. Bahlmann (2009), 'The Role of the Posterior Superior Temporal Cortex in Sentence Comprehension', *Neuroreport*, 20 (6): 563–8

Friederici, A. D. and J. Weissenborn (2007), 'Mapping Sentence Form Onto Meaning: The Syntax-Semantic Interface', *Brain Research*, 1,146: 50–8

Friedman, D. P., E. A. Murray, B. O'Neill and M. Mishkin (1986), 'Cortical Connections of the Somatosensory Fields of the Lateral Sulcus of Macaques: Evidence for a Corticolimbic Pathway for Touch', *Journal of Comparative Neurology*, 252: 323–47

Frijling, J. L. (2017), 'Preventing PTSD with Oxytocin: Effects of Oxytocin Administration on Fear Neurocircuitry and PTSD Symptom Development in Recently Trauma-Exposed Individuals', *European Journal of Psychotraumatology*, 8 (1): 1,302,652

Friston, K. J., P. Jezzard and R. Turner (1994), 'Analysis of Functional MRI Time-Series', *Human Brain Mapping*, 1: 153–71.

Frith, C. D. (2007), *Making Up the Mind: How the Brain Creates Our Mental World*, Oxford: Blackwell

Frith, C. D., K. Friston, P. F. Liddle and R. S. J. Frakowiak (1991), 'Willed Action and the Prefrontal Cortex in Man: A Study with PET', *Proceedings of the Royal Society of London. Series B. Containing Papers of a Biological Character*, 244: 241–6

Frith, C. D. and U. Frith (2005), 'Theory of Mind', *Current Biology*, 15 (17): R644–5, 209

Frith, U. (2008), *Autism: A Very Short Introduction*, Oxford: Oxford University Press

Froese, T., H. Iizuka and T. Ikegami (2014), 'Embodied Social Interaction Constitutes Social Cognition in Pairs of Humans: A Minimalist Virtual Reality Experiment', *Scientific Reports*, 4: 3,672

Fromkin, V. and R. Rodman (1997), *An Introduction to Language*, 6th edn, New York, NY: Harcourt Brace Jovanovich

Frost, R., B. C. Armstrong, N. Siegelman and M. H. Christiansen (2015), 'Domain Generality Versus Modality Specificity: The Paradox of Statistical Learning', *Trends in Cognitive Sciences*, 19 (3): 117–25

Fulkerson, M. (2014), 'Rethinking the Senses and Their Interactions: The Case for Sensory Pluralism', *Frontiers in Psychology*, 145: article 1,426

Funahashi, S., C. J. Bruce and P. S. Goldman-Rakic (1993), 'Dorsolateral Prefrontal Lesions and Oculomotor Delayed-Response Performance: Evidence for Mnemonic "Scotomas"', *Journal of Neuroscience*, 13: 1,479–97

Gabora, L. and A. Russon (2011), 'The Evolution of Intelligence', in *The Cambridge Handbook of Intelligence* (pp. 328–42), Cambridge: Cambridge University Press

Galaburda, A. M. (1994), 'Developmental Dyslexia and Animal Studies: At the Interface Between Cognition and Neurology', *Cognition*, 50: 133–49

Gallagher, H. L., F. Happé, N. Brunswick et al. (2000), 'Reading the Mind in Cartoons and Stories: An fMRI Study of "Theory of Mind" in Verbal and Nonverbal Tasks', *Neuropsychologia*, 38: 11–21

Gallagher, S. (ed.), (2011), *The Oxford Handbook of the Self*, Oxford: Oxford University Press

Gallup, G. G. (1970), 'Chimpanzees: Self-recognition', *Science*, 167: 86–7

Gandhi, S. P., D. J. Heeger and G. M. Boynton (1999), Spatial Attention Affects Brain Activity in Human Primary Visual Cortex', *Proceedings of the National Academy of Sciences of the United States of America*, 96: 3,314–19

Gandour, J., D. Wong, L. Hsieh et al. (2000), 'A Crosslinguistic PET Study of Tone Perception', *Journal of Cognitive Neuroscience* 12: 207–22

Garden, G. A. and A. R. La Spada (2012), 'Intercellular (Mis)communication in Neurodegenerative Disease', *Neuron*, 73 (5): 886–901

Garden, G. A. and T. Moller (2006), 'Microglia Biology in Health and Disease', *Journal of Neuroimmune Pharmacology*, 1 (2), 2006: 127–37

Gardner, H., H. Brownell, W. Wapner and D. Michelow (1983), 'Missing the Point: The Role of the Right Hemisphere in the Processing of Complex Linguistic Materials', in E. Perecman (ed.), *Cognitive Processes in the Right Hemisphere* (pp. 169–92), New York, NY: Academic Press

Garrison, K. A., J. F. Santoyo, J. H. Davis et al. (2013), 'Effortless Awareness: Using Real-Rime Neurofeedback to Investigate Correlates of Posterior Cingulate Cortex Activity in Meditators' Self-Report', *Frontiers in Human Neuroscience*, 7: article 440

Gaugler, T., L. L. Klei, S. J. Sanders et al. (2014), 'Most Genetic Risk for Autism Resides with Common Variation', *Nature Genetics*, 46 (8): 881–5

Gendron, M., K. A. Lindquist, L. W. Barsalou and L. F. Barrett (2012), 'Emotion Words Shape Emotion Percepts', *Emotion*, 12 (2): 314–25

Geschwind, D. H. (2011), 'Genetics of Autism Spectrum Disorders', *Trends in Cognitive Science*, 15: 409–16

Geschwind, D. H. and J. Flint (2015), 'Genetics and Genomics of Psychiatric Disease', *Science*, 349 (6,255): 1,489–94

Geschwind, D. H. and M. W. State (2015), 'Gene Hunting in Autism Spectrum Disorder: On the Path to Precision Medicine', *Lancet Neurology*, 14 (11), 1,109–20

Geschwind, N. (1965), 'Disconnexion Syndromes in Animals and Man', *Brain*, 88: 585–644

—— (1970), 'The Organization of Language and the Brain', *Science*, 170: 940–4

Ghosh, H. S. (2019), 'Adult Neurogenesis and the Promise of Adult Neural Stem Cells', *Journal of Experimental Neuroscience*, 13: 1179069519856876

Gil-Ranedo, J., E. Gonzaga, K. J. Jaworek et al. (2019), 'STRIPAK Members Orchestrate Hippo and Insulin Receptor Signaling to Promote Neural Stem Cell Reactivation', *Cell Reports*, 27 (10): 2,921–33 e2925

Gilbert, C. D. and W. Li (2013), 'Top-Down Influences on Visual Processing'. *Nature Reviews Neuroscience*, 14 (5): 350–63

Glasser, M. F. and J. K. Rilling (2008), 'DTI Tractography of the Human Brain's Language Pathways', *Cerebral Cortex*, 18 (11): 2,471–82

Goldberg, I. I., M. Harel and R. Malach (2006), 'When the Brain Loses Its Self: Prefrontal Inactivation During Sensorimotor Processing', *Neuron*, 50 (2): 329–39

Goldstein, B. (2014), *Cognitive Psychology: Connecting Mind, Research, and Everyday Experience*, Belmont, CA: Wadsworth Publishing

Gomez, J. M., M. Verdu, A. Gonzalez-Megias and M. Mendez (2016), 'The Phylogenetic Roots of Human Lethal Violence', *Nature*, 538 (7,624): 233–7

Goodglass, H. (1993), *Understanding Aphasia*, San Diego, CA: Academic Press

Gopnik, A., A. N. Meltzoff and P. K. Kuhl (2001), *The Scientist in the Crib: What Early Learning Tells Us About the Mind*, New York, NY: Harper Collins

Gorno-Tempini, M. L., A. E. Hillis, S. Weintraub et al. (2011), 'Classification of Primary Progressive Aphasia and Its Variants', *Neurology* 76: 1,006–14

Gottfredson, L. (1998), 'The General Intelligence Factor', *Scientific American Presents* 9 (4): 24–9

Grandin, T. and R. Panek (2014), *The Autistic Brain*, Rider

Gratten, J., N. R. Wray, M. C. Keller and P. M. Visscher (2014), 'Large-Scale Genomics Unveils the Genetic Architecture of Psychiatric Disorders', *Nature Neuroscience*, 17 (6): 782–90

Graziano, M. (2016), 'Consciousness Engineered', *Journal of Consciousness Studies*, 23 (11–12): 98–115

Graziano, M. S. and S. Kastner (2011), 'Human Consciousness and Its Relationship to Social Neuroscience: A Novel Hypothesis', *Cognitive Neuroscience*, 2 (2): 98–113

Graziano, M. S. A. and T. W. Webb (2014), 'A Mechanistic Theory of Consciousness', *International Journal of Machine Consciousness*, 6: 163–76

Greenfield, S. (2000), *Brain Story: Why Do We Think and Feel As We Do?*, BBC Books

—— (2015), *A Day in the Life of the Brain: The Neuroscience of Consciousness from Dawn Till Dusk*, Penguin

Gregory, R. L. (2004), *The Oxford Companion to the Mind*, Oxford: Oxford University Press

Groopman, E. E., R. N. Carmody and R. W. Wrangham (2015), 'Cooking Increases Net Energy Gain from a Lipid-Rich Food', *American Journal of Physical Anthropolology*, 156 (1): 11–18

Gross, C. T. and N. S. Canteras (2012), 'The Many Paths to Fear', *Nature Reviews Neuroscience*, 13 (9): 651–8

Gross, James J. (2015), 'Emotion Regulation: Current Status and Future Prospects', *Psychological Inquiry*, 26 (1): 1–26

Guan, Z., M. Giustetto, S. Lomvardas et al. (2002), 'Integration of Long-Term–Memory-Related Synaptic Plasticity Involves Bidirectional Regulation of Gene Expression and Chromatin Structure', *Cell*, 111: 483–93

Guerra-Doce, E. (2015), 'Psychoactive Substances in Prehistoric Times: Examining the Archaeological Evidence', *Journal of Archaeology, Consciousness and Culture*, 8 (1): 91–112

Haddon, M. (2004), *The Curious Incident of the Dog in the Night-Time*, Vintage

Haftig, T. and E. Moser (2005), 'Microstructure of the Spatial Map in the Entorhinal Complex', *Nature*, 436: 801–8

Haggard, P. (2008), 'Human Volition: Towards a Neuroscience of Will', *Nature Reviews Neuroscience*, 9: 934–46

Haggard, P. and B. Libet. (2001), 'Conscious Intention and Brain Activity', *Journal of Consciousness Studies*, 8: 47–63

Hagoort, P. (2003), 'Interplay Between Syntax and Semantics During Sentence Comprehension: ERP Effects of Combining Syntactic and Semantic Violations', *Journal of Cognitive Neuroscience*, 15 (6): 883–99

Hall, D. A., I. S. Johnsrude, M. P. Haggard et al. (2002), 'Spectral and Temporal Processing in Human Auditory Cortex', *Cerebral Cortex*, 12 (2): 140–9

Hallett, P. J., O. Cooper, D. Sadi et al. (2014), 'Long-Term Health of Dopaminergic Neuron Transplants in Parkinson's Disease Patients', *Cell Reports*, 7 (6), 1,755–61

Hameroff, S. and R. Penrose (2014), 'Consciousness in the Universe: A Review of the "Orch OR" Theory', *Physics of Life Review*, 11: 39–112

Happé, F., S. Ehlers, P. Fletcher et al. (1996), '"Theory of Mind" in the Brain: Evidence from a PET Scan Study of Asperger Syndrome', *Neuroreport*, 8: 197–201

Happé, F. and U. Frith (eds) (2010), *Autism and Talent*, Oxford: Oxford University Press (first published in 2009 as a special issue of *Philosophical Transactions of the Royal Society of London: B. Biological Sciences*, vol. 364)

Hardt, O., E. O. Einarsson and K. Nader (2010), 'A Bridge Over Troubled Water: Reconsolidation as a Link Between Cognitive and Neuroscientific Memory Research Traditions', *Annual Review of Psychology*, 61: 141–67

Harnad, S. (2007), 'Can a Machine Be Conscious? How?', *Journal of Consciousness Studies*, 10 (4–5): 67–75

Harris, A. (2019), *Conscious: A Brief Guide to the Fundamental Mystery of the Mind*, Harper

Harris, P. (2011), 'BrainGate Gives Paralysed the Power of Mind Control', *Guardian*, 17 April 2011, www.theguardian.com/science/2011/apr/17/brain-implant-paralysis-movement

Harris, S. (2012), *Free Will*, New York, NY: Free Press

—— (2014), *Waking Up: A Guide to Spirituality Without Religion*, Bantam Press

Harrison, N. A., M. A. Gray, P. J. Gianaros and H. D. Critchley (2010), 'The Embodiment of Emotional Feelings in the Brain', *Journal of Neuroscience*, 30: 12,878–84

Hart, H. and K. Rubia (2012), 'Neuroimaging of Child Abuse: A Critical Review', *Frontiers in Human Neuroscience*, 6 (52): 1–24

Hauser, M., N. Chomsky and T. Fitch (2002), 'The Faculty of Language: What Is It, Who Has It, and How Did It Evolve?', *Science*, 298: 1,569–79

Hawkins, R. D., E. R. Kandel and C. H. Bailey (2006), 'Molecular Mechanisms of Memory Storage in Aplysia', *Biological Bulletin*, 210: 174–91

Hawrylycz, M., J. A. Miller, V. Menon et al. (2015), 'Canonical Genetic Signatures of the Adult Human Brain', *Nature Neuroscience*, 18 (12): 1,832–44

Hazlett, H. C., M. Poe, G. Gerig et al. (2005), 'Magnetic Resonance Imaging and Head Circumference Study of Brain Size in Autism: Birth Through Age 2 Years', *Archives of General Psychiatry*, 62: 1,366–76

Heimer, L. (1994), *The Human Brain and Spinal Cord: Functional Neuroanatomy and Dissection Guide*, 2nd edn, New York, NY: Springer

Henrich, J. and J. Broesch (2011), 'On the Nature of Cultural Transmission Networks: Evidence from Fijian Villages for Adaptive Learning Biases',

Philosophical Transactions of the Royal Society of London: B. Biological Sciences, 366 (1,567): 1,139–48

Heyes, C. (2018), *Cognitive Gadgets: The Cultural Evolution of Thinking*, Cambridge, MA: Belknap Press

Hibar, D. P., J. L. Stein, M. E. Renteria et al. (2015), 'Common Genetic Variants Influence Human Subcortical Brain Structures', *Nature* 520 (7,546): 224–9

Hickok, G. and D. Poeppel (2004), 'Dorsal and Ventral Streams: A Framework for Understanding Aspects of the Functional Anatomy of Language', *Cognition*, 92 (1–2): 67–99

—— (2007), 'The Cortical Organization of Speech Processing', *Nature Reviews Neuroscience*, 8: 393–402

Hidaka, B. H. (2012), 'Depression as a Disease of Modernity: Explanations for Increasing Prevalence', *Journal of Affective Disorders*, 140 (3): 205–14

Higashida, N. (2013), *The Reason I Jump: The Inner Voice of a Thirteen-Year-Old Boy with Autism*, New York, NY: Random House

Hill, E. (2004), 'Executive Dysfunction in Autism', *Trends in Cognitive Science*, 8: 26–32

Hinton, G. E. and S. J. Nolan (1987), 'How Learning Can Guide Evolution', *Complex Systems* 1: 495–502

Hirst, W., J. K. Yamashiro and A. Coman (2018), 'Collective Memory from a Psychological Perspective', *Trends in Cognitive Science*, 22 (5): 438–51

Hobbes, T. (1651), *Leviathan*, Penguin Classics, 2016

Hofstadter, D. R. (2007), 'A Fleeting Encounter with Zombies and Dualism [excerpt]', in D. R. Hofstadter, *I Am a Strange Loop* (pp. 342–9), New York, NY: Basic Books

Holland, O. (2007), 'A Strongly Embodied Approach to Machine Consciousness', *Journal of Consciousness Studies*, 14: 97–110

Holland, P. C. and M. Gallagher (2004), 'Amygdala-Frontal Interactions and Reward Expectancy', *Current Opinion in Neurobiology*, 14: 148–55

Holmes, J., K. A. Hilton, M. Place et al. (2014), 'Children with Low Working Memory and Children with ADHD: Same or Different?', *Frontiers in Human Neuroscience*, 8: 976

Horner, A. J., J. A. Bisby, E. Zotow et al. (2016), 'Grid-Like Processing of Imagined Navigation', *Current Biology*, 26 (6): 842–7

Horner, V., J. D. Carter, M. Suchak and F. B. de Waal (2011), 'Spontaneous Prosocial Choice by Chimpanzees', *Proceedings of the National Academy of Sciences of the United States of America*, 108 (33): 13,847–51

Hrdy, S. B. (2011), *Mothers and Others: The Evolutionary Origins of Mutual Understanding*, Cambridge, MA: Belknap Press

Huang, Y.-Y., X.-C. Li and E. R. Kandel. (1994), 'cAMP Contributes to Mossy Fiber LTP by Initiating Both a Covalently-Mediated Early Phase and a Macromolecular Synthesis-Dependent Late Phase', *Cell*, 79: 69–79

Huang, Y.-Y., K. C. Marti and E. R. Kandel (2000), 'Both Protein Kinase A and Mitogen-Activated Protein Kinase Are Required in the Amygdala for the Macromolecular Synthesis-Dependent Late Phase of Long-Term Potentiation', *Journal of Neuroscience*, 20: 6,317–25

Hübener, M. and T. Bonhoeffer (2010), 'Searching for Engrams', *Neuron*, 67: 363–71

Huck, M., E. Fernandez-Duque, P. Babb and T. Schurr (2014), 'Correlates of Genetic Monogamy in Socially Monogamous Mammals: Insights from Azara's Owl Monkeys, *Proceedings of the Royal Society of London. Series B, Containing Papers of a Biological Character*, 281 (1,782): 20140195

Hume, D. (1739), *A Treatise of Human Nature: Being an Attempt to Introduce the Experimental Method of Reasoning into Moral Subjects*, Penguin Classics, 1985

Humphrey, N. (1992), *A History of the Mind: Evolution and the Birth of Consciousness*, Göttingen: Copernicus

—— (2000), How to Solve the Mind-Body Problem' *Journal of Consciousness Studies*, 7: 5–112

—— (2016), 'Redder Than Red Illusionism or Phenomenal Surrealism?', *Journal of Consciousness Studies*, 23 (11–12): 116–23

Humphreys, L. G. (1979), 'The Construct of General Intelligence', *Intelligence*, 3 (2): 105–20

Huttunen, A. W., G. K. Adams and M. L. Platt (2018), 'Can Self-Awareness Be Taught? Monkeys Pass the Mirror Test – Again', *Proceedings of the National Academy of Sciences of the United States of America*, 114 (13): 3,281–3

Hyvärinen, J. and A. Poranen (1978), 'Movement-Sensitive and Direction and Orientation-Selective Cutaneous Receptive Fields in the Hand Area of the Post-Central Gyrus in Monkeys', *Journal of Physiology*, 283: 523–37

Iacoboni, M., I. Molnar-Szakacs, V. Gallese et al. (2005), 'Grasping the Intentions of Others with One's Own Mirror Neuron System', *PLoS Biology*, 3 (3): 529–35

Imada, T., Y. Zhang, M. Cheour et al. (2006), 'Infant Speech Perception Activates Broca's Area: A Developmental Magnetoencephalography Study', *Neuroreport*, 17: 957–62

Immordino-Yang, M. H., A. McColl, H. Damasio and A. R. Damasio (2009), 'Neural Correlates of Admiration and Compassion', *Proceedings of the National Academy of Sciences of the United States of America*, 106: 8,021–6

Immordino-Yang, M. H., J. A. Christodoulou and V. Singh (2012), 'Rest is Not Idleness: Implications of the Brain's Default Mode for Human Development and Education', *Perspectives on Psychological Science*, 7 (4): 352–64

Inzlicht, M., B. D. Bartholow and J. B. Hirsh (2015), 'Emotional Foundations of Cognitive Control', *Trends in Cognitive Sciences*, 19 (3): 126–32

Iverson, P., P. K. Kuhl, R. Akahane-Yamada et al. (2003), 'A Perceptual Interference Account of Acquisition Difficulties for Non-Native Phonemes', *Cognition*, 87: B47–57

Jackson, J. H. (1915), 'On Affections of Speech from Diseases of the Brain', *Brain*, 38: 107–74

Jacobs, N. S., T. A. Allen, N. Nguyen and N. J. Fortin (2013), 'Critical Role of the Hippocampus in Memory for Elapsed Time', *Journal of Neuroscience*, 33 (34): 13,888–93

Jacobsen, C. F. (1935), 'Function of the Frontal Association Area in Primates', *Archives of Neurology and Psychiatry*, 33: 558–69

Jamain, S., H. Quach, C. Betancur et al. (2003), 'Mutations of the X-Linked Genes Encoding Neuroligins NLGN3 and NLGN4 are Associated with Autism', *Nature Genetics*, 34: 27–9

James, W. (1884), 'What is an Emotion?', *Mind*, 9 (34): 189–90

—— (1890), 'The Consciousness of Self', in W. James, *The Principles of Psychology* (vol. 1, ch. 10, pp. 291–401), Macmillan

—— (1890), *The Principles of Psychology*, New York, NY: Dover, 1950

Jamieson, J. P., M. K. Nock and W. B. Mendes (2012), 'Mind Over Matter: Reappraising Arousal Improves Cardiovascular and Cognitive Responses to Stress', *Journal of Experimental Psychology: General* 141 (3): 417–22

—— (2013), 'Changing the Conceptualization of Stress in Social Anxiety Disorder Affective and Physiological Consequences', *Clinical Psychological Science*, 1: 363–74

Jeannerod, M. (1986), 'The Formation of Finger Grip During Prehension: A Cortically Mediated Visuomotor Pattern', *Behavioural Brain Research*, 19: 99–116

Jebelli, J. D., C. Hooper, G. A. Garden and J. M. Pocock (2012), 'Emerging Roles of p53 in Glial Cell Function in Health and Disease', *Glia*, 60 (4): 515–25

Jebelli, J. D., C. Hooper and J. M. Pocock (2014), 'Microglial p53 Activation is Detrimental to Neuronal Synapses During Activation-Induced Inflammation: Implications for Neurodegeneration', *Neuroscience Letters*, 583: 92–7

Jebelli, J. D., W. Su, S. Hopkins et al. (2015), 'Glia: Guardians, Gluttons, or Guides for the Maintenance of Neuronal Connectivity?' *Annals of the New York Academy of Sciences*, 1,351: 1–10

Jenkins, W. M., M. M. Merzenich, M. T. Ochs et al. (1990), 'Functional Reorganization of Primary Somatosensory Cortex in Adult Owl Monkeys After Behaviorally Controlled Tactile Stimulation', *Journal of Neurophysiology*, 63: 83–104

Jin, P., R. S. Alisch and S. T. Warren (2004), 'RNA, and MicroRNA in Fragile X Syndrome', *Nature Cell Biology*, 6: 1,048–53

Johanson, D. C. (2004), 'Lucy, Thirty Years Later: An Expanded View of Australopithecus afarensis', *Journal of Anthropological Research*, 60 (4): 465–86

Johanson, D. C. and M. A. Edey (1981), *Lucy: The Beginnings of Humankind*, New York, NY: Simon & Schuster

John-Henderson, N. A., M. L. Rheinschmidt and R. Mendoza-Denton (2015), 'Cytokine Responses and Math Performance: The Role of Stereotype Threat and Anxiety Reappraisals'. *Journal of Experimental Social Psychology*, 56: 203–6

Johnson, B. R. and S. K. Lam (2010), 'Self-Organization, Natural Selection, and Evolution: Cellular Hardware and Genetic Software', *BioScience*, 60 (11): 879–85

Johnson, J. and E. Newport (1989), 'Critical Period Effects in Sound Language Learning: The Influence of Maturation State on the Acquisition of English as a Second Language', *Cognitive Psychology*, 21: 60–99

Jones, E. G. (1986), 'Connectivity of the Primate Sensory-Motor Cortex', in E. G. Jones and A. Peters (eds), *Cerebral Cortex*, vol. 5, ch. 4: 'Sensory -Motor Areas and Aspects of Cortical Connectivity' (pp. 113–83), New York, NY/London: Plenum

Jonides, J., R. L. Lewis, D. E. Nee et al. (2008), 'The Mind and Brain of Short-Term Memory', *Annual Review of Psychology*, 59: 193–224

Joyce, J. (1922), *Ulysses*, Penguin Classics, 2015

Jung, R. (1974), Neuropsychologie und Neurophysiologie des Kontur und Formensehens in Zeichnerei und Malerei', in H. H. Wieck (ed.) *Psycho -pathologie Musischer Bestaltungen* (pp. 29–88), Stuttgart: Schaltauer

Jusczyk, P. W., A. D. Friederici, J. M. I. Wessels et al. (1993), 'Infants' Sensitivity to the Sound Patterns of Native Language Words', *Journal of Memory and Language*, 32: 402–20

Juskalian, R. (2020), 'A New Implant for Blind People Jacks Directly into the Brain', *MIT Technology Review*, www.technologyreview.com/s/615148/a-new-implant-for-blind-people-jacks-directly-into-the-brain/

Kaas, J. H. (2006), 'Evolution of the Meocortex', *Current Biology*, 16: R910–14

Kaas, J. H. and T. A. Hackett (1999) '"What" and "Where" Processing in Auditory Cortex', *Nature Neuroscience*, 2: 1,045–7

Kaas, J. H., R. J. Nelson, M. Sur et al. (1979), 'Multiple Representations of the Body Within the Primary Somatosensory Cortex of Primates', *Science*, 204: 521–3

Kaas, J. H., H. X. Qi, M. J. Burish et al. (2008), 'Cortical and Subcortical Plasticity in the Brains of Humans, Primates, and Rats After Damage to Sensory Afferents in the Dorsal Columns of the Spinal Cord', *Experimental Neurology*, 209: 407–16

Kahneman, D. (2012), *Thinking, Fast and Slow*, Penguin

Kaiser, J. (2014), '"Rejuvenation Factor" in Blood Turns Back the Clock in Old Mice', *Science*, 344 (6,184): 570–1

Kana, R. K., T. A. Keller, V. L. Cherkassky et al. (2009), 'Atypical Frontal-Posterior Synchronization of Theory of Mind Regions in Autism During Mental State Attribution', *Social Neuroscience*, 4: 135–52

Kandel, E. R. (2001), 'The Molecular Biology of Memory Storage: A Dialogue Between Genes and Synapses', *Science*, 294: 1,030–8

—— (2001), 'The Molecular Biology of Memory Storage: A Dialog Between Genes and Synapses' (Nobel Lecture), *Bioscience Reports*, 21: 565–611

—— (2006), *In Search of Memory: The Emergence of a New Science of Mind*, New York, NY: W. W. Norton

Kandel, E. R., J. H. Schwartz, T. M. Jessell et al. (2012), *Principles of Neural Science*, 5th edn, New York, NY: McGraw-Hill Education/ Medical.

Kanwisher, N. (2001), 'Neural Events and Perceptual Awareness', *Cognition*, 79: 89–113

Karni, A., G. Meyer, C. Rey-Hipolito et al. (1998), 'The Acquisition of Skilled Motor Performance: Fast and Slow Experience-Driven Changes in Primary Motor Cortex', *Proceedings of the National Academy of Sciences of the United States of America*, 95: 861–8

Kashdan, T. B. and A. S. Farmer (2014), 'Differentiating Emotions Across Contexts: Comparing Adults With and Without Social Anxiety Disorder Using Random, Social Interaction, and Daily Experience Sampling', *Emotion*, 14 (3): 629–38

Kastner, S. and L. G. Ungerleider (2000), 'Mechanisms of Visual Attention in the Human Cortex', *Annual Review of Neuroscience*, 23: 315–41

Kathirvel, N. and A. Mortimer (2013), 'Causes, Diagnosis and Treatment of Visceral Hallucinations', *Progress in Neurology and Psychiatry*, January/February, 6–10

Katz, L. F., A. C. Maliken and N. M. Stettler (2012), 'Parental Meta-Emotion Philosophy: A Review of Research and Theoretical Framework', *Child Development Perspectives*, 6 (4): 417–22

Kaufman, S. B. (2016), talk given at the Bay Area Discovery Museum's Creativity Forum, Fairmont San Francisco Hotel, San Francisco, CA

Keleman, K., S. Krüttner, M. Alenius and B. J. Dickson (2007), 'Function of the Drosophila CPEB Protein Orb2 in Long-Term Courtship Memory', *Nature Neuroscience*, 10: 1,587–93

Kelly, S. D., C. Kravitz and M. Hopkins (2004), 'Neural Correlates of Bimodal Speech and Gesture Comprehension', *Brain and Language*, 89 (1): 253–60

Kerchner, G. A and R. A. Nicoll (2008), 'Silent Synapses and the Emergence of a Postsynaptic Mechanism for LTP', *Nature Reviews Neuroscience*, 9: 813–25

Kessels, H. W. and R. Malinow (2009), 'Synaptic AMPA Receptor Plasticity and Behavior', *Neuron*, 61: 340–50

Key, B. (2016), 'Why Fish Do Not Feel Pain', *Animal Sentience: An Interdisciplinary Journal on Animal Feeling*, 1 (3): 39

Kihlstrom, J. F. (1985), 'Hypnosis', *Annual Review of Psychology*, 36: 385–418

Kinno, R., M. Kawamura, S. Shioda and K. L. Sakai (2008), 'Neural Correlates of Noncanonical Syntactic Processing Revealed by a Picture-Sentence Matching Task', *Human Brain Mapping*, 29 (9): 1,015–27

Kirk, R. (2015), 'Zombies', in E. N. Zalta (ed.), *The Stanford Encyclopedia of Philosophy*, Summer, http://plato.stanford.edu/archives/sum2015/entries/zombies/

Kirk, R. and J. E. R. Squires (1974), 'Zombies vs Materialists', *Proceedings of the Aristotelian Society*, 48: 135–52

Klin, A., W. Jones, R. Schultz et al. (2002), 'Defining and Quantifying the Social Phenotype in Autism', *American Journal of Psychiatry*, 159: 895–908

Knudsen, E. I. (2004), 'Sensitive Periods in the Development of the Brain and Behavior', *Journal of Cognitive Neuroscience*, 16: 1,412–25

Kobatake, E., G. Wang and K. Tanaka (1998), 'Effects of Shape-Discrimination Training on the Selectivity of Inferotemporal Cells in Adult Monkeys', *Journal of Neurophysiology*, 80: 324–30

Koch, C., M. Massimini, M. Boly and G. Tononi (2016), 'Neural Correlates of Consciousness: Progress and Problems', *Nature Reviews Neuroscience*, 17: 307–21

Koenigs, M., L. Young, R. Adolphs et al. (2007), 'Damage to the Prefrontal Cortex Increases Utilitarian Moral Judgments', *Nature*, 446: 908–11

Kolb, B. and I. Q. Whishaw (1990), *Fundamentals of Human Neuropsychology*, 3rd edn, New York, NY: Freeman

Komatsu, H. (2006), 'The Neural Mechanisms of Perceptual Filling-In', *Nature Reviews Neuroscience*, 7: 220–31

Komers, P. and P. N. M. Brotherton (1997), 'Female Space Use is the Best Predictor of Monogamy in Mammals', *Proceedings of the Royal Society of London. Series B, Containing Papers of a Biological Character*, 264: 1,261–70

Kovács, Á. M., E. Téglás and A. D. Endress (2010), 'The Social Sense: Susceptibility to Others' Beliefs in Human Infants and Adults', *Science*, 330: 1,830–4

Kringelbach, M. L. and K. C. Berridge (eds) (2010), *Pleasures of the Brain*, New York, NY: Oxford University Press

Kühl, H. S., A. K. Kalan, M. Arandjelovic et al. (2016), 'Chimpanzee Accumulative Stone Throwing', *Scientific Reports*, 6: 22,219

Kuhl, P. K. (2000), 'A New View of Language Acquisition', *Proceedings of the National Academy of Sciences of the United States of America*, 97: 11,850–7

—— (2004), 'Early Language Acquisition: Cracking the Speech Code', *Nature Reviews Neuroscience*, 5: 831–43

Kuhl, P. K., J. Andruski, I. Christovich et al. (1997), 'Cross-Language Analysis of Phonetic Units in Language Addressed to Infants', *Science*, 277: 684–6

Kuhl, P. K. and M. Rivera-Gaxiola (2008), 'Neural Substrates of Language Acquisition', *Annual Review of Neuroscience*, 31: 511–34

Kuhl, P. K., F.-M. Tsao and H.-M. Liu (2003), 'Foreign-Language Experience in Infancy: Effects of Short-Term Exposure and Social Interaction on Phonetic Learning', *Proceedings of the National Academy of Sciences of the United States of America*, 100: 9,096–101

Kuhl, P. K., K. A. Williams, F. Lacerda et al. (1992), 'Linguistic Experience

Alters Phonetic Perception in Infants by 6 Months of Age', *Science*, 255: 606–8

Kuhnen, C. M. and B. Knutson (2005), 'The Neural Basis of Financial Risk Taking', *Neuron*, 47: 763–70

Kukushkin, N. V. and T. J. Carew (2017), 'Memory Takes Time', *Neuron*, 95 (2): 259–79

Kumar, S., K. E. Stephan, J. D. Warren et al. (2007), 'Hierarchical Processing of Auditory Objects in Humans', *PLoS Computational Biology*, 3 (6): e100, 977–85

Kutas, M. and K. D. Federmeier (2000), 'Electrophysiology Reveals Semantic Memory Use in Language Comprehension', *Trends in Cognitive Sciences*, 4 (12): 463–70

LaBar, K. S. and R. Cabeza (2006), 'Cognitive Neuroscience of Emotional Memory', *Nature Reviews Neuroscience*, 7: 54–64

LaBar, K. S., J. C. Gatenby, J. C. Gore et al. (1998), 'Human Amygdala Activation During Conditioned Fear Acquisition and Extinction: A Mixed Trial fMRI Study', *Neuron*, 20: 937–45

LaBar, K. S., J. E. LeDoux, D. D. Spencer and E. A. Phelps (1995), 'Impaired Fear Conditioning Following Unilateral Temporal Lobectomy in Humans', *Journal of Neuroscience*, 15: 6,846–55

Labuda, D., J. F. Lefebvre, P. Nadeau and M. H. Roy-Gagnon (2010), Female -to-Male Breeding Ratio in Modern Humans: An Analysis Based on Historical Recombinations', *American Journal of Human Genetics*, 86 (3): 353–63

Lai, C. S., S. E. Fisher, J. A. Hurst et al. (2001), 'A Forkhead-Domain Gene is Mutated in a Severe Speech and Language Disorder', *Nature*, 413: 519–23

Lam, M., J. W. Trampush, J. Yu et al. (2017), 'Large-Scale Cognitive GWAS Meta-Analysis Reveals Tissue-Specific Neural Expression and Potential Nootropic Drug Targets', *Cell Reports*, 21 (9): 2,597–613

Lancaster, M. A., M. Renner, C. A. Martin (2013), 'Cerebral Organoids Model Human Brain Development and Microcephaly', *Nature*, 501 (7,467): 373–9

Langdell, T. (1978), 'Recognition of Faces: An Approach to the Study of Autism', *Journal of Child Psychology and Psychiatry*, 19 (3): 255–68

Leavens, D. A., K. A. Bard and W. D. Hopkins (2019), 'The Mismeasure of Ape Social Cognition', *Animal Cognition*, 22 (4): 487–504

Leaver, A. M. and J. P. Rauschecker (2010), 'Cortical Representation of

Natural Complex Sounds: Effects of Acoustic Features and Auditory Object Category', *Journal of Neuroscience*, 30 (22): 7,604–12

LeBihan, D. (2003), 'Looking into the Functional Architecture of the Brain with Diffusion MRI', *Nature Reviews Neuroscience*, 4: 469–80

Lebon, G. (2008), *The Crowd: A Study of the Popular Mind*, Digireads.com

Lebrecht, S., M. Bar, L. F. Barrett and M. J. Tarr (2012), 'Micro-Valences: Perceiving Affective Valence in Everyday Objects', *Frontiers in Perception Science*, 3 (107): 1–5

LeDoux, J. E. (1996), *The Emotional Brain*, New York, NY: Simon & Schuster

—— (2000), 'Emotion Circuits in the Brain', *Annual Review of Neuroscience*, 23: 155–84

—— (2003), *Synaptic Self: How Our Brains Become Who We Are*, Penguin

Lee, R. B. (1969), 'Eating Christmas in the Kalahari', http://people.morrisville.edu/~reymers/readings/ANTH101/EatingChristmas-Lee.pdf

Lee, S. H., R. Blake and D. J. Heeger (2005), 'Traveling Waves of Activity in Primary Visual Cortex During Binocular Rivalry', *Nature Neuroscience*, 8: 22–3

—— (2007), 'Hierarchy of Cortical Responses Underlying Binocular Rivalry', *Nature Neuroscience* 10: 1,048–54

Lek, M., K. J. Karczewski, E. V. Minikel et al. (2016), 'Analysis of Protein-Coding Genetic Variation in 60,706 Humans', *Nature*, 536 (7,616): 285–91

Lenneberg, E. (1967), *Biological Foundations of Language*, New York, NY: Wiley

Lennie, P. (2003), 'The Cost of Cortical Computation', *Current Biology*, 13: 493–7

Leppänen, J. M. and C. A. Nelson (2009), 'Tuning the Developing Brain to Social Signals of Emotions', *Nature Reviews Neuroscience*, 10 (1): 37–47

Lesser, R. P., S. Arroyo, J. Hart and B. Gordon (1994), 'Use of Subdural Electrodes for the Study of Language Functions', in A. Kertesz (ed.), *Localization and Neuro-Imaging in Neuropsychology* (pp. 57–72), San Diego, CA: Academic Press

Lewis, A. G., J.-M. Schoffelen, H. Schriefers and M. Bastiaansen (2016), 'A Predictive Coding Perspective on Beta Oscillations During Sentence -Level Language Comprehension', *Frontiers in Human Neuroscience*, 10: 85

Libet, B. (1982), 'Brain Stimulation in the Study of Neuronal Functions for Conscious Sensory Experiences', *Human Neurobiology*, 1: 235–242

—— (2004), *Mind Time: The Temporal Factor in Consciousness*, Cambridge, MA: Harvard University Press

Libet, B., C. A. Gleason, E. W. Wright and D. K. Pearl (1983), 'Time of Conscious Intention to Act in Relation to Onset of Cerebral Activity (Readiness-Potential): The Unconscious Initiation of a Freely Voluntary Act', *Brain*, 106 (pt 3): 623–42

Lieberman, M. (2015), *Social: Why Our Brains Are Wired to Connect*, Oxford: Oxford University Press

Lieberman, P. (2017), 'Comment on "Monkey Vocal Tracts Are Speech-Ready"', *Science Advances* 3 (7): e1700442

Lin, D., M. P. Boyle, P. Dollar et al. (2011), 'Functional Identification of an Aggression locus in the Mouse Hypothalamus', *Nature*, 470: 221–6

Linebarger, M., M. Schwartz and E. Saffran (1983), 'Sensitivity to Grammatical Structure in So-Called Agrammatic Aphasics', *Cognition*, 13: 361–92

Liu, H.-M., P. K. Kuhl and F.-M. Tsao (2003), 'An Association Between Mothers' Speech Clarity and Infants' Speech Discrimination Skills', *Developmental Science*, 6: F1–10

Liu, L., J. Lei, S. J. Sanders et al. (2014), 'DAWN: A Framework to Identify Autism Genes and Subnetworks Using Gene Expression and Genetics', *Molecular Autism*, 5 (1): 22

Logothetis, N. K., J. Pauls, M. Augath et al. (2001), 'Neurophysiological Investigation of the Basis of the fMRI Signal', *Nature*, 412: 150–7

Lømo, T. (2003), 'The Discovery of Long-Term Potentiation', *Philosophical Transactions of the Royal Society of London: B. Biological Sciences*, 358 (1,432): 617–20

Ludlow, P., Y. Nagasawa and D. Stoljar (eds) (2004), *There's Something About Mary: Essays on Phenomenal Consciousness and Frank Jackson's Knowledge Argument*, Cambridge, MA: MIT Press

Lumer, E. D., K. J. Friston and G. Rees (1998), 'Neural Correlates of Perceptual Rivalry in the Human Brain', *Science*, 280: 1,930–4

Lundqvist, L. and H. Lindner (2017), 'Is the Autism-Spectrum Quotient a Valid Measure of Traits Associated with the Autism Spectrum? A Rasch Validation in Adults With and Without Autism Spectrum Disorders', *Journal of Autism and Developmental Disorders*, 47 (7): 2,080–91

Luria, A. (1980), *Higher Cortical Functions in Man*, New York, NY: Basic Books

Lutz, W. and E. Kebede (2018), 'Education and Health: Redrawing the Preston Curve', *Population and Development Review*, 44 (2): 343–61

McDonald, R. J. and N. M. White (1993), 'A Triple Dissociation of Memory Systems: Hippocampus, Amygdala, and Dorsal Striatum', *Behavioural Neuroscience*, 107: 3–22

McFarland, D. J., J. Daly, C. Boulay and M. Parvaz (2017), 'Therapeutic Applications of BCI Technologies', *Brain Computer Interfaces* (Abingdon), 47 (1–2): 37–52

McGaugh, J. L. (2003), *Memory and Emotions: The Making of Lasting Memories* New York, NY: Columbia University Press

McGaugh, J. L., L. Cahill and B. Roozendaal (1996), 'Involvement of the Amygdala in Memory Storage: Interaction with Other Brain Systems', *Proceedings of the National Academy of Sciences of the United States of America*, 93: 13,508–14

McHugh, T. J., M. W. Jones, J. J. Quinn et al. (2007), 'Dentate Gyrus NMDA Receptors Mediate Rapid Pattern Separation In The Hippocampal Network', *Science*, 317: 94–9

McKenzie, A. L., S. S. Nagarajan, T. P. Roberts et al. (2003), 'Somatosensory Representation of the Digits and Clinical Performance in Patients with Focal Hand Dystonia', *American Journal of Physical Medicine and Rehabilitation*, 82: 737–49

McKinney, S. M., M. Sieniek, V. Godbole et al. (2020), 'International Evaluation of an AI System for Breast Cancer Screening', *Nature*, 577 (7,788): 89–94

MacLean, P. D. (1990), *The Triune Brain in Evolution*, New York, NY: Plenum

McMenamin, B. W., S. J. E. Langeslag, M. Sirbu et al. (2014), 'Network Organization Unfolds Over Time During Periods of Anxious Anticipation', *Journal of Neuroscience*, 34 (34): 11,261–73

Macphail, E. M. (2009), 'Evolution of Consciousness', in T. Bayne, A. Cleeremans and P. Wilken (eds), *The Oxford Companion to Consciousness* (pp. 276–9), Oxford: Oxford University Press

Maddieson, I. (2015), talk given at the 170th Meeting of the Acoustical Society of America (ASA), quoted in E. Underwood, 'Human Language May Be Shaped Climate and Terrain', www.sciencemag.org/news/2015/11/human-language-may-be-shaped-climate-and-terrain

Mahr, J. and G. Csibra (2017), 'Why Do We Remember? The Communicative Function of Episodic Memory', *Behavioral and Brain Sciences*, 1–93

Malaspina, D., S. Harlap, S. Fenniget et al. (2001), 'Advancing Paternal Age and the Risk of Schizophrenia', *Archives of General Psychiatry*, 58: 361–7

Malaspina, D., A. Reichenberg, M. Weiser et al. (2005), 'Paternal Age and Intelligence: Implications for Age-Related Genomic Changes in Male Germ Cells', *Psychiatric Genetics*, 15: 117–25

Malinowski, P. (2013), 'Neural Mechanisms of Attentional Control in Mindfulness Meditation', *Frontiers in Neuroscience*, 7: article 8

Manning, J. (2016), 'Come to Think of It . . . or Not: Dartmouth Study Shows How Memories Can be Intentionally Forgotten', www.dartmouth.edu/press-releases/memories-can-be-intentionally-forgotten-050516.html

Maoz, H., H. Z. Gvirts, M. Sheffer and Y. Bloch (2017), 'Theory of Mind and Empathy in Children with ADHD', *Journal of Attention Disorders*, 1087054717710766

Marchini, J. (2016), A Reference Panel of 64,976 Haplotypes for Genotype Imputation', *Nature Genetics*, 48 (10): 1,279–83

Marcus, G. F., S. Vijayan, S. Bandi Rao and P. M. Vishton (1999), 'Rule Learning by Seven-Month-Old Infants', *Science*, 283 (5,398): 77–80

Marder, E. and A. L. Taylor (2011), 'Multiple Models to Capture the Variability in Biological Neurons and Networks', *Nature Neuroscience*, 14: 133–8

Marek, S., J. S. Siegel, E. M. Gordon et al. (2018), 'Spatial and Temporal Organization of the Individual Human Cerebellum', *Neuron*, 100 (4): 977–93 e977

Maren, S. (1999), 'Long-Term Potentiation in the Amygdala: A Mechanism for Emotional Learning and Memory', *Trends in Neuroscience*, 22: 561–7

Maren, S. (2005), 'Synaptic Mechanisms of Associative Memory in the Amygdala', *Neuron*, 47: 783–6

Marien, P., H. Ackermann, M. Adamaszek et al. (2014), 'Consensus Paper: Language and the Cerebellum: An Ongoing Enigma', *Cerebellum*, 13 (3): 386–410

Maril, A., A. D. Wagner and D. L. Schacter (2001), 'On the Tip of the Tongue: An Event-Related fMRI Study of Semantic Retrieval Failure and Cognitive Conflict', *Neuron* 31: 653–60

Marshall, J. C. and P. W. Halligan (1995), 'Seeing the Forest but Only Half the Trees?', *Nature*, 373: 521–3

Marshall, W. H., C. N. Woolsey and P. Bard (1941), 'Observations on Cortical Somatic Sensory Mechanisms of Cat and Monkey', *Journal of Neurophysiology*, 4: 1–24

Martin, A. and L. L. Chao (2001), 'Semantic Memory and the Brain: Structure and Processes', *Current Opinion in Neurobiology*, 11: 194–201

Martin, K. C., A. Casadio, H. Zhu et al. (1997), 'Synapse-Specific, Long-Term Facilitation of Aplysia Sensory to Motor Synapses: A Function for Local Protein Synthesis in Memory Storage', *Cell*, 91: 927–38

Martinussen, R., J. Hayden, S. Hogg-Johnson and R. Tannock (2005), 'A Meta-Analysis of Working Memory Impairments in Children with Attention-Deficit/Hyperactivity Disorder', *Journal of the American Academy of Child and Adolescent Psychiatry*, 44 (4): 377–84

Mayberg, H. S., M. Liotti, S. K. Brannan et al. (1999), 'Reciprocal Limbic -Cortical Function and Negative Mood: Converging PET Findings in Depression and Normal Sadness', *American Journal of Psychiatry*, 156: 675–82

Mayer, A., and B. E. Träuble (2013), 'Synchrony in the Onset of Mental State Understanding Across Cultures? A Study Among Children in Samoa', *International Journal of Behavioural Development*, 37: 21–8

Mazoyer, B. M., N. Tzourio, V. Frak et al. (1993), 'The Cortical Representation of Speech', *Journal of Cognitive Neuroscience*, 5: 467–79

Medina, J. F., C. J. Repa, M. D. Mauk and J. E. LeDoux (2002), 'Parallels Between Cerebellum- and Amygdala-Dependent Conditioning', *Nature Reviews Neuroscience*, 3: 122–31

Mesulam, M.-M. (1985), *Principles of Behavioral Neurology*, Philadelphia, PA: F. A. Davis

Metzinger, T. (2009), *The Ego Tunnel: The Science of the Mind and the Myth of the Self*, New York, NY: Basic Books

Meyer, L., J. Obleser and A. D. Friederici (2013), 'Left Parietal Alpha Enhancement During Working Memory-Intensive Sentence Processing', *Cortex*, 49 (3): 711–21

Mikulas, W. L. (2007), 'Buddhism & Western Psychology: Fundamentals of Integration', *Journal of Consciousness Studies*, 14 (4), 4–49

Miller, E. K. and J. D. Cohen (2001), 'An Integrative Theory of Prefrontal Cortex Function', *Annual Review of Neuroscience*, 24: 167–202

Milner, B. and M. Petrides (1984), 'Behavioural Effects of Frontal-Lobe Lesions in Man', *Trends in Neuroscience*, 7: 403–7

Mishkin, M. and J. Aggleton (1981), 'Multiple Functional Contributions of the Amygdala in the Monkey', in: Y. Ben-Ari (ed.), *The Amygdaloid Complex* (pp. 409–20), Amsterdam: Elsevier/North Holland

Miyawaki, K., W. Strange, R. Verbrugge et al. (1975), 'An Effect of Linguistic

Experience: The Discrimination of [R] And [L] by Native Speakers of Japanese and English', *Perception and Psychophysics*, 18: 331–40

Mizokawa, A. and S. Lecce (2016), 'Sensitivity to Criticism and Theory of Mind: A Cross Cultural Study on Japanese and Italian Children', *European Journal of Developmental Psychology*, 14 (2): 159–71

Mlot, S. (2018), 'World Go Champion Hopes to Reclaim Title in AI Rematch', www.geek.com/tech/world-go-champ-hopes-to-reclaim-title-in-ai-rematch-1726687/

Mogilner, A., J. A. Grossman, V. Ribraly et al. (1993), 'Somato-Sensory Cortical Plasticity in Adult Humans Revealed by Magneto-Encephalography', *Proceedings of the National Academy of Sciences of the United States of America*, 9: 3,593–7

Moisse, K. (2017), 'Many Children with Autism Get Significantly Smarter Over Time', *Spectrum News*, www.spectrumnews.org/news/many-children-autism-get-significantly-smarter-time/

Moll, J., F. Kruegcr, R. Zahn et al. (2006), 'Human Fronto-Mesolimbic Networks Guide Decisions About Charitable Donation', *Proceedings of the National Academy of Sciences of the United States of America*, 103: 15,623–8

Molnar, Z., G. J. Clowry, N. Sestan et al. (2019), 'New Insights into the Development of the Human Cerebral Cortex', *Journal of Anatomy*, 235 (3): 432–51

Monod, J. (1971), *Chance and Necessity: An Essay on the Natural Philosophy of Modern Biology*, Vintage

Morris, J. S., C. D. Frith, D. I. Perrett et al. (1996), 'A Different Neural Response in the Human Amygdala to Fearful and Happy Facial Expressions', *Nature*, 383: 812–15

Moses, D. A., M. K. Leonard, J. G. Makin and E. F. Chang (2019), 'Real-Time Decoding of Question-and-Answer Speech Dialogue Using Human Cortical Activity', *Nature Communications*, 10 (1): 3,096

Motta, S. C., M. Goto, F. V. Gouveia et al. (2009), 'Dissecting the Brain's Fear System Reveals the Hypothalamus is Critical for Responding in Subordinate Conspecific Intruders', *Proceedings of the National Academy of Sciences of the United States of America*, 106: 4,870–5

Mottron, L., M. Dawson, I. Soulières et al. (2006), 'Enhanced Perceptual Functioning in Autism: An Update, and Eight Principles of Autistic Perception', *Journal of Autism and Developmental Disorders*, 36 (1): 27–43

Mountcastle, V. B. (1984), 'Central Nervous Mechanisms in Mechanoreceptive

Sensibility', in I. Darian-Smith (ed.), *Handbook of Physiology*, sect. 1, vol. 3, pt 2 (pp. 789–878), Bethesda, MD: American Physiological Society

Mozolic, J. L., S. Hayasaka and P. J. Laurienti (2010), 'A Cognitive Training Intervention Increases Resting Cerebral Blood Flow in Healthy Older Adults', *Frontiers in Human Neuroscience*, 4: 16

Mueller, J. L., A. D. Friederici and C. Männel (2012), 'Auditory Perception at the Root of Language Learning', *Proceedings of the National Academy of Sciences of the United States of America*, 109 (39): 15,953–8

Murata, A., V. Gallese, G. Luppino et al. (2000), 'Selectivity for the Shape, Size, and Orientation of Objects for Grasping in Neurons of Monkey Parietal Area AIP', *Journal of Neurophysiology*, 83: 2,580–601

Murray, E. A., S. P. Wise and K. S. Graham (2017), *The Evolution of Memory Systems: Ancestors, Anatomy, and Adaptations* (Oxford Psychology Series), Oxford: Oxford University Press

Murty, M. R. and V. K. Murty (2013), *The Mathematical Legacy of Srinivasa Ramanujan*, New York, NY: Springer

'Muskie' (2002), Institute for the Study of the Neurotypical, http://erikengdahl.se/autism/isnt/

Naab, P. J. and J. A. Russell (2007), 'Judgments of Emotion from Spontaneous Facial Expressions of New Guineans', *Emotion* 7 (4): 736–44

Nadler, A., C. F. Camerer, D. T. Zava et al. (2019), 'Does Testosterone Impair Men's Cognitive Empathy? Evidence from Two Large-Scale Randomized Controlled Trials', *Proceedings of the Royal Society B. Biological Sciences*, 286 (1,910)

Nagel, T. (1974), 'What Is It Like to Be a Bat?', *Philosophical Review*, 83: 435–50.

Nakazawa, K., M. C. Quirk, R. A. Chitwood et al. (2002), 'Requirement for Hippocampal CA3 NMDA Receptors in Associative Memory Recall', *Science*, 297: 211–18

Nakazawa, K., L. D. Sun, M. C. Quirk et al. (2003), 'Hippocampal CA3 NMDA Receptors Are Crucial for Memory Acquisition of One-Time Experience', *Neuron*, 38: 306–15

Naqvi, N. H., D. Rudrauf, H. Damasio and A. Bechara, (2007), 'Damage to the Insula Disrupts Addiction to Cigarette Smoking', *Science*, 315: 531–4

Naya, Y., M. Yoshida and Y. Miyashita (2001), 'Backward Spreading of Memory-Related Signal in the Primate Temporal Cortex', *Science*, 291: 661–4

Nelson, E., C. Rolian, L. Cashmore and S. Shultz (2011), 'Digit Ratios Predict Polygyny in Early Apes, Ardipithecus, Neanderthals and Early

Modern Humans but Not in Australopithecus', *Proceedings of the Royal Society of London. Series B, Containing Papers of a Biological Character*, 278 (1,711): 1,556–63

Neubauer, S., J. J. Hublin and P. Gunz (2018), 'The Evolution of Modern Human Brain Shape', *Science Advances*, 4 (1): eaao5961

Neubert, F.-X., R. B. Mars, A. G. Thomas et al. (2014), 'Comparison of Human Ventral Frontal Cortex Areas for Cognitive Control and Language with Areas in Monkey Frontal Cortex', *Neuron*, 81 (3): 700–13

Neville, H. J., S. A. Coffey, D. Lawson et al. (1997), 'Neural Systems Mediating American Sign Language: Effects of Sensory Experience and Age of Acquisition', *Brain and Language*, 57: 285–308

Newberry, M. G., C. A. Ahern, R. Clark and J. B. Plotkin (2017), 'Detecting Evolutionary Forces in Language Change', *Nature*, 551 (7,679): 223–6

Newman, J. (1956), *The World of Mathematics*, vol. 1, New York, NY: Dover

Newport, E. L. and R. N. Aslin (2004), 'Learning at a Distance I: Statistical Learning of Non-Adjacent Dependencies', *Cognitive Psychology*, 48: 127–62

Nicholls, R. E., J. M. Alarcon, G. Malleret et al. (2008), 'Transgenic Mice Lacking NMDAR-Dependent LTD Exhibit Deficits in Behavioral Flexibility', *Neuron*, 58: 104–17

Nickl-Jockschat, T., U. Habel, T. M. Michel et al. (2012), 'Brain Structure Anomalies in Autism Spectrum Disorder: A Meta-Analysis of VBM Studies Using Anatomic Likelihood Estimation', *Human Brain Mapping*, 33(6): 1,470–89

Nieuwenhuys, R., J. Voogd and C. Huijzen (1988), *The Human Central Nervous System: A Synopsis and Atlas*, 3rd edn, Berlin: Springer-Verlag

Noble, D. (2008), *The Music of Life: Biology Beyond Genes*, Oxford: Oxford University Press

Noble, K. G., S. M. Houston, N. H. Brito et al. (2015), 'Family Income, Parental Education and Brain Structure in Children and Adolescents', *Nature Neuroscience*, 18 (5): 773–8

Noë, A. (ed.) (2002), 'Is the Visual World a Grand Illusion?', special issue, *Journal of Consciousness Studies*, 9 (5–6)

Nunn, C. (ed.) (2009), 'Defining Consciousness', special issue, *Journal of Consciousness Studies*, 16 (5)

Nyberg, L., R. Habib, A. R. McIntosh and E. Tulving (2000), 'Reactivation of Encoding-Related Brain Activity During Memory Retrieval', *Proceedings of the National Academy of Sciences of the United States of America*, 97: 11,120–4

O'Donnell, M. B., J. B. Bayer, C. N. Cascio and E. B. Falk (2017), 'Neural Bases of Recommendations Differ According to Social Network Structure', *Social Cognitive and Affective Neuroscience*, 12 (1): 61–9

Ohman, A. (2005), 'The Role of the Amygdala in Human Fear: Automatic Detection of Threat', *Psychoneuroendocrinology*, 10: 953–8

O'Keefe, J. and J. Dostrovsky (1971), 'The Hippocampus as a Spatial Map: Preliminary Evidence from Unit Activity in the Freely-Moving Rat', *Brain Research*, 34: 171–5

Oldendorf, W. H. (1980), *The Quest for an Image of the Brain: Computerized Tomography in the Perspective of Past and Future Imaging Methods*, New York, NY: Raven

Olivares, F. A., E. Vargas, C. Fuentes et al. (2015), 'Neurophenomenology Revisited: Second-Person Methods for the Study of Human Consciousness', *Frontiers in Psychology*, 6: 673

Opendak, M. and E. Gould (2015), 'Adult Neurogenesis: A Substrate for Experience-Dependent Change', *Trends in Cognitive Sciences*, 19 (3): 151–61

Opitz, B. and A. D. Friederici (2003), 'Interactions of the Hippocampal System and the Prefrontal Cortex in Learning Language-Like Rules', *NeuroImage*, 19 (4): 1,730–7

O'Regan, J. K. and A. Noë (2001), 'A Sensorimotor Account of Vision and Visual Consciousness', *Behavioral and Brain Sciences*, 24 (5), 883–1,031

O'Roak, B. J. and M. W. State (2008), 'Autism Genetics: Strategies, Challenges, and Opportunities', *Autism Research*, 1 (1), 4–17

O'Roak, B. J., L. Vives, S. Girirajan et al. (2012), 'Sporadic Autism Exomes Reveal a Highly Interconnected Protein Network of de novo Mutations', *Nature*, 485 (7,397), 246–50

Ostler, N. (2018), 'Have We Reached Peak English in the World?', *Guardian*, 27 February 2018, www.theguardian.com/commentisfree/2018/feb/27/reached-peak-english-britain-china

Ozonoff, S., A. M. Iosif, F. Baguio et al. (2010), 'A Prospective Study of the Emergence of Early Behavioral Signs of Autism', *Journal of the American Academy of Child and Adolescent Psychiatry*, 49: 256–66

Ozonoff, S., S. Macari, G. S. Young et al. (2008), 'Atypical Object Exploration at 12 Months of Age is Associated with Autism in a Prospective Sample', *Autism*, 12: 457–72

Packard, M. G., R. Hirsh and N. M. White (1989), 'Differential Effects of Fornix and Caudate Nucleus Lesions on Two Radial Maze Tasks: Evidence for Multiple Memory Systems', *Journal of Neuroscience*, 9: 1,465–72

Padmanabhan, A., C. J. Lynch, M. Schaer and V. Menon (2017), 'The Default Mode Network in Autism', *Biological Psychiatry: Cognitive Neuroscience and Neuroimaging*, 2 (6): 476–86

Pagel, M. (2017), 'Darwinian Perspectives on the Evolution of Human Languages', *Psychonomic Bulletin and Review*, 24 (1): 151–7

Panksepp, J. (1998), *Affective Neuroscience: The Foundations of Human and Animal Emotions*, New York, NY: Oxford University Press

Panksepp, J. and J. B. Panksepp (2013), 'Toward a Cross-Species Understanding of Empathy', *Trends in Neurosciences*, 36 (8): 489–96

Papineau, D. (2003), 'Confusions About Consciousness', *Richmond Journal of Philosophy*, 5

Pardini, D. A., A. Raine, K. Erickson and R. Loeber (2014), 'Lower Amygdala Volume in Men is Associated with Childhood Aggression, Early Psychopathic Traits, and Future Violence', *Biological Psychiatry*, 75 (1): 73–80

Pare, D. (2003), 'Role of the Basolateral Amygdala in Memory Consolidation', *Progress in Neurobiology*, 70: 409–20

Parfit, D. (1987), 'Divided Minds and the Nature of Persons', in C. Blakemore and S. Greenfield (eds), *Mindwaves* (pp. 19–26), Oxford: Blackwell

Parikshak, N. N., R. Luo, A. Zhang et al. (2013), 'Integrative Functional Genomic Analyses Implicate Specific Molecular Pathways and Circuits in Autism', *Cell*, 155 (5), 1,008–21

Paton, J. J., M. A. Belova, S. E. Morrison and C. D. Salzman (2006), 'The Primate Amygdala Represents the Positive and Negative Value of Visual Stimuli During Learning', *Nature*, 439: 865–70

Peechakara, B. V. and M. Gupta (2020), 'Vitamin B3', Treasure Island, FL: StatPearls

Pelphrey, K. A., S. Shultz, C. M. Hudac and B. C. Vander Wyk (2011), 'Research Review: Constraining Heterogeneity: The Social Brain and Its Development in Autism Spectrum Disorder', *Journal of Child Psychology and Psychiatry*, 52: 631–4.

Penfield, W. and E. Boldrey (1937), 'Somatic Motor and Sensory Representation in the Cerebral Cortex of Man as Studied by Electrical Stimulation', *Brain*, 60: 389–443

Penfield, W. and T. Rasmussen (1950), *The Cerebral Cortex of Man*, New York, NY: Macmillan

Penfield, W. and L. Roberts (1959), *Speech and Brain Mechanisms*, Princeton, NJ: Princeton University Press

Pennisi, E. (2004), 'Human Evolution. The Primate Bite: Brawn Versus Brain?', *Science*, 303 (5,666): 1,957

Perani, D., E. Paulesu, N. S. Galles et al. (1998) 'The Bilingual Brain: Proficiency and Age of Acquisition of the Second Language', *Brain*, 121 (10): 1,841–52

Perenin, M. T. and A. Vighetto (1988), 'Optic Ataxia: A Specific Disruption in Visuo-Motor Mechanisms, I: Different Aspects of the Deficit in Reaching for Objects', *Brain*, 111 (pt 3): 643–74

Peterson, S. E., P. T. Fox, M. I. Posner et al. (1988), 'Positron Emission Tomographic Studies of the Cortical Anatomy of Single-Word Processing', *Nature*, 331: 585–9

Peterson, J. (2017), 'Personality and Its Transformations: Lecture 18: Biology and Traits-Openness, Intelligence & Creativity', https://www.youtube.com/watch?v=D7Kn5p7TP_Y

Petersson, K.-M., C. Forkstam and M. Ingvar (2004), 'Artificial Syntactic Violations Activate Broca's Region', *Cognitive Science*, 28 (3): 383–407

Petitmengin, C. and J. P. Lachaux (2013), 'Microcognitive Science: Bridging Experiential and Neuronal Microdynamics', *Frontiers in Human Neuroscience*, 7: article 617

Petkov, C. I. and E. D. Jarvis (2012), 'Birds, Primates, and Spoken Language Origins: Behavioral Phenotypes and Neurobiological Substrates', *Frontiers in Evolutionary Neuroscience*, 4: 12

Petrides, M. (1994), 'Frontal Lobes and Behavior', *Current Opinion in Neurobiology*, 4: 207–11

Petrovich, G. D. (2011), 'Learning and the Motivation to Eat: Forebrain Circuitry', *Physiology and Behavior*, 104: 582–9

Pettito, L. A., S. Holowka, L. E. Sergio et al. (2004), 'Baby Hands That Move to the Rhythm of Language: Hearing Babies Acquiring Sign Language Babble Silently on the Hands', *Cognition*, 93: 43–73

Phelps, E. A. (2006), 'Emotion and Cognition: Insights from Studies of the Human Amygdala', *Annual Review of Psychology*, 57: 27–53

Phelps, E. A. and J. E. LeDoux (2005), 'Contributions of the Amygdala to Emotion Processing: From Animal Models to Human Behavior', *Neuron*, 48: 175–87

Piet, J. and E. Hougaard (2011), 'The Effect of Mindfulness-Based Cognitive Therapy for Prevention of Relapse in Recurrent Major Depressive Disorder: A Systematic Review and Meta-Analysis', *Clinical Psychology Review*, 31, 1,032–40

Pinker, S. (1994), *The Language Instinct*, New York, NY: William Morrow

—— (1997), *How the Mind Works*, New York, NY: W. W. Norton

—— (2002), *The Blank Slate: The Modern Denial of Human Nature*, New York, NY: Viking

Pisotta, I. and M. Molinari (2014), 'Cerebellar Contribution to Feedforward Control of Locomotion', *Frontiers in Human Neuroscience*, 8: 1–5

Pitkänen, A., V. Savander and J. E. LeDoux (1997), 'Organization of Intra-Amygdaloid Circuitries in the Rat: An Emerging Framework for Understanding Functions of the Amygdala', *Trends in Neuroscience*, 20: 517–23

Pittenger, C., S. Fasano, D. Mazzocchi-Jones et al. (2006), 'Impaired bidirectional Synaptic Plasticity and Procedural Memory Formation in Striatum-Specific cAMP Response Element-Binding Protein-Deficient Mice', *Journal of Neuroscience*, 261: 2,808–13

Pivot, S. (2003), 'La Commune, les Communards, les écrivains ou la haine et la gloire', *La Revue de l'association des anciens élèves de l'École Nationale d'Administration, Politique et Littérature*, December 2003

Poldrack, R. A., J. Clark, E. J. Pare-Blagoev (2001), 'Interactive Memory Systems in the Human Brain', *Nature*, 414: 546–50

Pons, T. P., P. E. Garraghty, D. P. Friedmanet al. (1987), 'Physiological Evidence for Serial Processing in Somatosensory Cortex', *Science*, 237: 417–20

Posner, M. (1994), 'Attention: The Mechanisms of Consciousness', *Proceedings of the National Academy of Sciences of the United States of America*, 91: 7,398–403

Posner, M. I. (1980), 'Orienting of Attention', *Quarterly Journal of Experimental Psychology*, 32: 3–25

Posner, M. I. and S. Dahaene (1994), 'Attentional Networks', *Trends in Neuroscience*, 17: 75–9

Power, R., D. Salazar-Garcia, L. Straus et al. (2015), 'Microremains from El Mirón Cave Human Dental Calculus Suggest a Mixed Plant–Animal Subsistence Economy During the Magdalenian in Northern Iberia', *Journal of Archaeological Science*, 60: 39–46

Price, A. R., M. F. Bonner, J. E. Peelle, and M. Grossman (2015), 'Converging Evidence for the Neuroanatomic Basis of Combinatorial Semantics in the Angular Gyrus', *Journal of Neuroscience*, 35 (7): 3,276–84

Price, C. J. (2010), 'The Anatomy of Language: A Review of 100 fMRI Studies Published in 2009', *Annals of the New York Academy of Sciences*, 1,191: 62–88

Price, D. D. and J. J. Barrell (2012), 'Developing a Science of Human Meanings and Consciousness', in D. D. Price and J. J. Barrell, *Inner Experience and Neuroscience: Merging Both Perspectives* (pp. 1–30), Cambridge, MA: MIT Press

Protzko, J. (2016), 'Does the Raising IQ-Raising g Distinction Explain the Fadeout Effect?', *Intelligence*, 56: 65–71

Qazi, R., A. M. Gomez, D. C. Castro et al. (2019), 'Wireless Optofluidic Brain Probes for Chronic Neuropharmacology and Photostimulation', *Nature Biomedical Engineering*, 3 (8): 655–69

Quirk, G. J. and D. R. Gehlert (2003), 'Inhibition of the Amygdala: Key to Pathological States?', *Annals of the New York Academy of Sciences*, 985: 263–72

Raichle, M. E. (2010), 'Two Views of Brain Function', *Trends in Cognitive Science*, 14 (4): 180–90

Raine, A. (2014), *The Anatomy of Violence: The Biological Roots of Crime*, Penguin

Ramachandran, V. S. (1993), 'Behavioral and Magnetoencephalographic Correlates of Plasticity in the Adult Human Brain', *Proceedings of the National Academy of Sciences of the United States of America*, 90: 10,413–20

Ramón y Cajal, S. (1995), *Histology of the Nervous System of Man and Vertebrates*, 2 vols, trans. N. Swanson and L. W. Swanson, New York, NY: Oxford University Press

Ramsden, S., F. M. Richardson, G. Josse et al. (2011), 'Verbal and Non-Verbal Intelligence Changes in the Teenage Brain', *Nature*, 479 (7,371): 113–16

Ransom, M., S. Fazelpour and C. Mole (2017), 'Attention in the Predictive Mind', *Consciousness and Cognition*, 47: 99–112

Rao, R. P., A. Stocco, M. Bryan et al. (2014), 'A Direct Brain-to-Brain Interface in Humans', *PLoS ONE*, 9 (11): e111332

Rao, S. C., G. Rainer and E. K. Miller (1997), 'Integration of What and Where in the Primate Prefrontal Cortex', *Science*, 276: 821–4

Rauch, S. L., L. M. Shin and E. A. Phelps (2006), 'Neurocircuitry Models of Posttraumatic Stress Disorder and Extinction: Human Neuroimaging Research – Past, Present, and Future', *Biological Psychiatry*, 60: 376–82

Rautiainen, M. R., T. Paunio, E. Repo-Tiihonen et al. (2016), 'Genome-Wide Association Study of Antisocial Personality Disorder', *Translational Psychiatry*, 6 (9): e883

Redcay, E. and E. Courchesne (2005), 'When is the Brain Enlarged in Autism? A Meta-Analysis of All Brain Size Reports', *Biological Psychiatry*, 58: 1–9

Redmond Jr, D. E., S. Weiss, J. D. Elsworth et al. (2010), 'Cellular Repair in the Parkinsonian Nonhuman Primate Brain', *Rejuvenation Research*, 13 (2–3): 188–94

Rees, G., G. Kreimanand, C. Koch (2002), 'Neural Correlates of Consciousness in Humans', *Nature Reviews Neuroscience*, 3: 261–70

Reid, T. (1785), *Essays on the Intellectual Powers of Man*, Edinburgh: John Bell

Reiss, D., L. Leve and J. Neiderhiser (2013), 'How Genes and the Social Environment Moderate Each Other', *American Journal of Public Health*, 103: S111–21

Rensink, R. A. and G. Kuhn (2015), 'A Framework for Using Magic to Study the Mind', *Frontiers in Psychology*, 5: article 1,508

Repa, J. C., J. Muller, J. Apergis et al. (2001), 'Two Different Lateral Amygdala Cell Populations Contribute to the Initiation and Storage of Memory', *Nature Neuroscience*, 4: 724–31

Ress, D., B. T. Backus and D. J. Heeger (2000), 'Activity in Primary Visual Cortex Predicts Performance in a Visual Detection Task', *Nature Neuroscience*, 3: 940–5

Reysen, M. B. (2007), 'The Effects of Social Pressure on False Memories', *Memory and Cognition*, 35 (1): 59–65

Rilling, J. K. and R. A. Seligman (2002), 'A Quantitative Morphometric Comparative Analysis of the Primate Temporal Lobe', *Journal of Human Evolution*, 42 (5): 505–33

Rizzolatti, G. and M. A. Arbib (1998), 'Language Within Our Grasp', *Trends in Neurosciences*, 21 (5): 188–94

Robinson, E. B., K. E. Samocha, J. A. Kosmicki et al. (2014), 'Autism Spectrum Disorder Severity Reflects the Average Contribution of de novo and Familial Influences', *Proceedings of the National Academy of Sciences*, 111 (42): 15,161–5

Robinson, E. B., B. St Pourcain, V. Anttila et al. (2016), 'Genetic Risk for Autism Spectrum Disorders and Neuropsychiatric Variation in the General Population', *Nature Genetics*, 48 (5), 552–5

Rochat, P. (2003), 'Five Levels of Self-Awareness as They Unfold Early in Life', *Consciousness and Cognition*, 12: 717–31

Rockland, K. S. and N. Ichinohe (2004), 'Some Thoughts on Cortical Minicolumns', *Experimental Brain Research*, 158: 265–77

Rodrigues, S. M., G. E. Schafe and J. E. LeDoux (2004), 'Molecular Mechanisms Underlying Emotional Learning and Memory in the Lateral Amygdala', *Neuron*, 44: 75–91

Rogan, M. T., K. S. Leon, D. L. Perez et al. (2005), 'Distinct Neural Signatures for Safety and Danger in the Amygdala and Striatum of the Mouse', *Neuron*, 46: 309–20

Rolls, E. (1999), *The Brain and Emotion*, New York, NY: Oxford University Press

Ronald, A. and R. A. Hoekstra (2011), 'Autism Spectrum Disorders and Autistic Traits: A Decade of New Twin Studies', *American Journal of Medical Genetics B Neuropsychiatric Genetics*, 156B: 255–74

Rondot, P., J. de Recondo and J. L. Dumas (1977), 'Visuomotor Ataxia', *Brain*, 100: 355–76

Rose Markus, H. and B. Schwartz (2010), 'Does Choice Mean Freedom and Well-Being?', *Journal of Consumer Research*, 37 (2): 344–55

Rosenberg, M. D., E. S. Finn, D. Scheinost et al. (2017), 'Characterizing Attention with Predictive Network Models', *Trends in Cognitive Sciences*, 21 (4): 290–302

Rosenthal, R. and L. Jacobson (1968), 'Pygmalion in the Classroom', *Urban Review*, 3 (1): 16–20

Rossi, S., M. F. Gugler, A. D. Friederici and A. Hahne (2006), 'The Impact of Proficiency on Syntactic Second Language Processing of German and Italian: Evidence from Event-Related Potentials', *Journal of Cognitive Neuroscience*, 18 (12): 2,030–48

Rottenberg, J. (2014), *The Depths: The Evolutionary Origins of the Depression Epidemic*, New York, NY: Basic Books

Rousseau, J.-J. (1782), *The Confessions of Jean-Jacques Rousseau*, Penguin Classics, 1973

Royal Society, The (2019), *iHuman: Blurring Lines Between Mind and Machine*, September 2019, Royal Society

Rubens, A. B. and D. F. Benson (1971), 'Associative Visual Agnosia', *Archives of Neurology*, 24: 305–16

Ruhlen, M. (2011), interviewed for Life's Little Mysteries, quoted at LiveScience, www.livescience.com/16541-original-human-language-yoda-sounded.html

Rumpel, S., J. LeDoux, A. Zador and R. Malinow (2005), 'Postsynaptic Receptor Trafficking Underlying a Form of Associative Learning', *Science*, 308: 83–8

Saffran, J. R., R. N. and E. L. Newport (1996) 'Statistical Learning by 8-Month-Old Infants', *Science*, 274: 1,926–8

Salter, M. W. and S. Beggs (2014), 'Sublime Microglia: Expanding Roles for the Guardians of the CNS', *Cell*, 158 (1): 15–24

Salzman, C. D., M. A. Belova and J. J. Paton (2005), 'Beetles, Boxes and Brain Cells: Neural Mechanisms Underlying Valuation and Learning', *Current Opinion in Neurobiology*, 6: 721–9

Sammler, D., S. A. Kotz, K. Eckstein et al. (2010), 'Prosody Meets Syntax: The Role of the Corpus Callosum', *Brain*, 133: 2,643–55

Sampson, R. J., J. H. Laub and C. Wimer (2006), 'Does Marriage Reduce Crime? A Counterfactual Approach to Within-Individual Causal Effects', *Criminology*, 44 (3): 465–508

Samson, D., I. A. Apperly, C. Chiavarino and G. W. Humphreys (2004), 'Left Temporoparietal Junction is Necessary for Representing Someone Else's Belief', *Nature Neuroscience*, 7: 499–500

Sanders, S. J., X. He, A. J. Willsey et al. (2015), 'Insights into Autism Spectrum Disorder Genomic Architecture and Biology from 71 Risk Loci', *Neuron*, 87 (6), 1,215–33

Sandin, S., P. Lichtenstein, R. Kuja-Halkola et al. (2014), 'The Familial Risk of Autism', *JAMA: Journal of the American Medical Association*, 311 (17), 1,770–7

Schacter, D. L. and D. R. Addis (2007), 'The Cognitive Neuroscience of Constructive Memory: Remembering the Past and Imagining the Future', *Philosophical Transactions of the Royal Society of London: B. Biological Sciences*, 362: 773–86

Schacter, D. L. and E. F. Loftus (2013), 'Memory and Law: What Can Cognitive Neuroscience Contribute?', *Nature Neuroscience*, 16 (2): 119–23

Scheele, D., N. Striepens, O. Güntürkün et al. (2012), 'Oxytocin Modulates Social Distance Between Males and Females', *Journal of Neuroscience*, 32 (46): 16,074–9

Schirmer, A. and S. A. Kotz (2006), 'Beyond the Right Hemisphere: Brain Mechanisms Mediating Vocal Emotional Processing', *Trends in Cognitive Sciences*, 10 (1): 24–30

Schneiderman, I., O. Zagoory-Sharon, J. F. Leckman and R. Feldman (2012), 'Oxytocin During the Initial Stages of Romantic Attachment: Relations to Couples' Interactive Reciprocity', *Psychoneuroendocrinology*, 37 (8): 1,277–85

Schumann, C. M., J. Hamstra, B. L. Goodlin-Jones et al. (2004). 'The Amygdala is Enlarged in Children but Not Adolescents with Autism; the Hippocampus is Enlarged at All Ages', *Journal of Neuroscience*, 24: 6,392–401

Schwartz, A. J., A. Boduroglu and A. H. Gutchess (2014), 'Cross-Cultural Differences in Categorical Memory Errors', *Cognitive Science*, 38 (5): 997–1,007

Scudellari, M. (2016). 'How iPS Cells Changed the World', *Nature*, 534 (7,607), 310–12

Sebat, J., B. Lakshmi, D. Malhotra et al. (2007), 'Strong Association of de novo Copy Number Variation with Autism', *Science*, 316: 445–9

Seidl, A. (2007), 'Infants' Use and Weighting of Prosodic Cues in Clause Segmentation', *Journal of Memory and Language*, 57 (1): 24–48

Selkoe, D. J., E. Mandelkow and D. M. Holtzman (2012), 'The Biology of Alzheimer Disease', *Cold Spring Harbor Perspectives in Medicine*, 2 (1)

Senju, A., V. Southgate, S. White and U. Frith (2009), 'Mindblind Eyes: An Absence of Spontaneous Theory of Mind in Asperger Syndrome', *Science*, 325: 883–5

Seymour, B., N. Daw, P., Dayan et al. (2007), 'Differential Encoding of Losses and Gains in the Human Striatum', *Journal of Neuroscience*, 27: 4,826–31

Shadlen, M. (1997), 'Look but Don't Touch or Vice Versa', *Nature*, 386: 122–3

Shadmehr, R., J. Brandt and S. Corkin (1998), 'Time-Dependent Motor Memory Processes in Amnesic Subjects', *Journal of Neurophysiology*, 80 (3): 1,590–7

Shahaeian, A., C. C. Peterson, V. Slaughter and H. M. Wellman (2011), 'Culture and the Sequence of Steps in Theory of Mind Development', *Developmental Psychology*, 47 (5): 1,239–47

Shen, G., T. Horikawa, K. Majima and Y. Kamitani (2019), 'Deep Image Reconstruction from Human Brain Activity', *PLoS Computational Biology*, 15 (1): e1006633

Sheskin, M. (2018), 'The Inequality Delusion: Why We've Got the Wealth Gap All Wrong', www.newscientist.com/article/mg23731710-300-the-inequality-delusion-why-weve-got-the-wealth-gap-all-wrong/

Si, K., M. Giustetto, A. Etkin et al. (2003), 'A Neuronal Isoform of CPEB Regulates Local Protein Synthesis and Stabilizes Synapse-Specific Long-Term Facilitation in Aplysia', *Cell*, 115: 893–904

Si, K., S. Lindquist and E. R. Kandel (2003), 'A Neuronal Isoform of the Aplysia CPEB Has Prion-Like Properties', *Cell*, 115: 879–91

Siegal, M. and R. Varley (2002), 'Neural Systems Involved in "Theory of Mind"', *Nature Reviews Neuroscience*, 3 (6), 463–71

Sigurdsson, T., V. Doyère, C. K. Cain and J. E. LeDoux (2007), 'Long-Term Potentiation in the Amygdala: A Cellular Mechanism of Fear Learning and Memory', *Neuropharmacology*, 52: 215–27

Silk, J. B. (2007), 'Social Components of Fitness in Primate Groups', *Science*, 317 (5,843): 1,347–51

Silva-Pereyra, J., M. Rivera-Gaxiola and P. K. Kuhl (2005), 'An Event-Related Brain Potential Study of Sentence Comprehension in Preschoolers: Semantic and Morphosyntatic Processing', *Cognitive Brain Research*, 23 (2–3): 247–58

Simons, D. J. and R. A. Rensink (2005), 'Change Blindness: Past, Present, And Future', *Trends in Cognitive Sciences*, 9 (1), 16–20

Sinclair, J. (1993), 'Don't Mourn for Us', www.autreat.com/dont_mourn.html

Singer, J. (2017), *Neurodiversity: The Birth of an Idea*, Judy Singer

Singer, P. (2011), *The Expanding Circle: Ethics, Evolution, and Moral Progress*, Princeton, NJ: Princeton University Press

Singer, T., B. Seymour, J. O'Doherty et al. (2004),' Empathy for Pain Involves the Affective but Not Sensory Components of Pain', *Science*, 303: 1,157–62

Skerry, A. E. and R. Saxe (2015), 'Neural Representations of Emotion Are Organized Around Abstract Event Features', *Current Biology*, 25 (15): 1,945–54

Skinner, B. F. (1938), *The Behavior of Organisms: An Experimental Analysis*, New York, NY: Appleton-Century-Crofts

—— (1957), *Verbal Behavior*, Acton, MA: Copely Publishing Group

Slager, R. E., T. L. Newton, C. N. Vlangos et al. (2003), 'Mutations in RAI1 Associated with Smith-Magenis Syndrome', *Nature Genetics*, 33: 1–3

Sloman, A. and R. Chrisley (2003), 'Virtual Machines and Consciousness', *Journal of Consciousness Studies*, 10 (4–5): 133–72

Smith, D., P. Schlaepfer, K. Major et al. (2017), 'Cooperation and the Evolution of Hunter-Gatherer Storytelling', *Nature Communications*, 8 (1): 1,853

Smith, K. S. and K. C. Berridge (2005), 'The Ventral Pallidum and Hedonic Reward: Neuromechanical Maps of Sucrose "Liking" and Food Intake', *Journal of Neuroscience*, 25: 8,637–49

Solomon, A. (2002), *The Noonday Demon: An Anatomy of Depression*, Vintage

Solomon, M., A. M. Iosif, V. P. Reinhardt et al. (2017), 'What Will My Child's Future Hold? Phenotypes of Intellectual Development in 2-8-Year-Olds with Autism Spectrum Disorder', *Autism Research*, 11 (1): 121–32

Soon, C. S., M. Brass, H. J. Heinze and J. D. Haynes (2008), 'Unconscious Determinants of Free Decisions in the Human Brain', *Nature Neuroscience*, 11 (5): 543–5

Spering, M. and M. Carrasco (2015), 'Acting Without Seeing: Eye Movements Reveal Visual Processing Without Awareness', *Trends in Neurosciences*, 38 (4): 247–58

Stedman, H. H., B. W. Kozyak, A. Nelson et al. (2004), 'Myosin Gene Mutation Correlates with Anatomical Changes in the Human Lineage', *Nature*, 428 (6,981): 415–18

Stellar, J. E., N. John-Henderson, C. L. Anderson et al. (2015), 'Positive Affect and Markers of Inflammation: Discrete Positive Emotions Predict Lower Levels of Inflammatory Cytokines', *Emotion*, 15 (2): 129–33

Sternberg, R. J., interviewed for *Reader's Digest*, https://www.rd.com/culture/practical-intelligence/

Sternberg, R. J. and S. B. Kaufman (2011), *The Cambridge Handbook of Intelligence,* Cambridge: Cambridge University Press

Steward, O. and E. M. Schuman (2001), 'Protein Synthesis at Synaptic Sites on Dendrites', *Annual Review of Neuroscience*, 24: 299–325

Stoddard, G. (1943), *The Meaning of Intelligence*, Macmillan

Strawson, G. (2006), 'Panpsychism?: Reply to Commentators with a Celebration of Descartes', *Journal of Consciousness Studies*, 13 (10–11): 184–280

Stromswold, K., D. Caplan, N. Alpert and S. Rauch (1996), 'Localization of Syntactic Comprehension Using Positron Emission Tomography', *Brain and Language* 52: 452–73

Suddendorf, T. and D. L. Butler (2013), 'The Nature of Visual Self-Recognition', *Trends in Cognitive Sciences*, 17 (3): 121–7

Swanson, L. W. and G. D. Petrovich (1998), 'What is the Amygdala?', *Trends in Neuroscience*, 21: 323–31

Swisher, K. and L. Goode (2017), 'Recode: Too Embarrassed to Ask', *Vox*, 30 March 2017, www.vox.com/2017/3/30/15130136/transcript-mary-lou-jepsen-one-laptop-per-child-too-embarrassed-to-ask-live-sxsw

Tabuchi, K., J. Blundell, M. H. Etherton et al. (2007), 'A Neuroligin-3 Mutation Implicated in Autism Increases Inhibitory Synaptic Transmission in Mice', *Science*, 318: 71–6

Tan, A. A. and D. L. Molfese (2009), 'ERP Correlates of Noun and Verb Processing in Preschool-Age Children', *Biological Psychology*, 8 (1): 46–51

Tang, Y.-Y., B. K. Hölzel and M. I. Posner (2015), 'The Neuroscience of Mindfulness Meditation', *Nature Reviews Neuroscience*, 16 (4): 213–25

Tatler, B. W. and M. F. Land (2011), 'Vision and the Representation of the Surroundings in Spatial Memory', *Philosophical Transactions of the Royal Society of London: B. Biological Sciences*, 366: 596–610

Thierry, G., M. Vihman and M. Roberts (2003), 'Familiar Words Capture the Attention of 11-Month-Olds in Less Than 250 ms', *Neuroreport*, 14 (18): 2,307–10

Thompson, E. and D. Zahavi (2007), 'Phenomenology', in P. D. Zelazo, M. Moskovitch and E. Thompson (eds), *The Cambridge Handbook of Consciousness* (pp. 67–87), Cambridge: Cambridge University Press

Thompson, K. G., K. L. Biscoe and T. R. Sato (2005), 'Neuronal Basis of Covert Spatial Attention in the Frontal Eye Field', *Journal of Neuroscience*, 25 (41), 9,479–87

Tian, B., D. Reser, A. Durham et al. (2001), 'Functional Specialization in Rhesus Monkey Auditory Cortex', *Science*, 292: 290–3

Tomasello, M. (2008), *Origins of Human Communication*, Cambridge, MA: MIT Press

Tong, F. (2003), 'Primary Visual Cortex and Visual Awareness', *Nature Reviews Neuroscience*, 4: 219–29

Tononi, G. (2015), 'Integrated Information Theory', *Scholarpedia*, 10 (1): 464, www.scholarpedia.org/ article/Integrated_information_theory

Treisman, A. M. and G. Gelade (1980), 'A Feature-Integration Theory of Attention', *Cognitive Psychology*, 12: 97–136

Tremblay, L. and W. Schultz (1999), 'Relative Reward Preference in Primate Orbitofrontal Cortex', *Nature*, 398: 704–8

Trojano, L. and D. Grossi (1998), '"Pure" Constructional Apraxia – A Cognitive Analysis of a Single Case', *Behavioural Neurology*, 11: 43–9

Tsao, F.-M., H.-M. Liu and P. K. Kuhl (2004), 'Speech Perception in Infancy Predicts Language Development in the Second Year of Life: A Longitudinal Study', *Child Development*, 75: 1,067–84

Tsuda, M., S. Beggs, M. W. Salter and K. Inoue (2013), 'Microglia and Intractable Chronic Pain', *Glia*, 61 (1): 55–61

Tulving, E. and D. L. Schacter (1990), 'Priming and Human Memory Systems', *Science*, 247: 301–6

Ullman, M. T. (2001), 'A Neurocognitive Perspective on Language: The Declarative/Procedural Model', *Nature Reviews Neuroscience*, 2 (10): 717–26

Ungerleider, L. G. and M. Mishkin (1982), 'Two Cortical Visual Systems', in D. J. Ingle, M. A. Goodale and R. J. W. Mansfield (eds), *Analysis of Visual Behavior* (pp. 549–86), Cambridge, MA: MIT Press

Vaina, L. M. (1994), 'Functional Segregation of Color and Motion Processing in the Human Visual Cortex: Clinical Evidence', *Cerebral Cortex*, 4: 555–72

Vallbo, Å. B., K. A. Olsson, K. G. Westberg and F. J. Clark (1984),

'Microstimulation of Single Tactile Afferents from the Human Hand: Sensory Attributes Related to Unit Type and Properties of Receptive Fields', *Brain*, 107: 727–49

Vandenberghe, R., A. C. Nobre and C. J. Price (2002), 'The Response of Left Temporal Cortex to Sentences', *Journal of Cognitive Neuroscience*, 14 (4): 550–60

Van Essen, D. C., H. A. Drury, J. Dickson et al. (2001), 'An Integrated Software Suite for Surface-Based Analyses of Cerebral Cortex', *Journal of the American Medical Informatics Association*, 8: 443–59

Vitebsky, P. (2005), *Reindeer People: Living with Animals and Spirits in Siberia*, HarperPerennial

Voss, U., K. Schermelleh-Engel, J. Windt et al. (2013), 'Measuring Consciousness in Dreams: The Lucidity and Consciousness in Dreams Scale', *Consciousness and Cognition*, 22: 8–21

Vuilleumier, P., J. L. Armony, K. Clarke et al. (2002), 'Neural Response to Emotional Faces With and Without Awareness: Event-Related fMRI in a Parietal Patient with Visual Extinction and Spatial Neglect', *Neuropsychologia*, 40: 2, 156–66

Wagner, A. D., E. J. Paré-Blagoev, J. Clark and R. A. Poldrack (2001), 'Recovering Meaning: Left Prefrontal Cortex Guides Control-Led Semantic Retrieval', *Neuron*, 31: 329–38

Walker, A. K., A. Kavelaars, C. J. Heijnen and R. Dantzer (2014), 'Neuroinflammation and Comorbidity of Pain and Depression', *Pharmacological Reviews*, 66 (1): 80–101

Walker, D. L., D. J. Toufexis and M. Davis (2003), 'Role of the Bed Nucleus of the Stria Terminalis Versus the Amygdala in Fear, Stress, and Anxiety', *European Journal of Pharmacology*, 463: 199–216

Wang, K., H. Zhang, D. Ma et al. (2009), 'Common Genetic Variants on 5p14.1 Associate with Autism Spectrum Disorders', *Nature*, 459 (7,246): 528–33

Wang, L., L. Uhrig, B. Jarraya and S. Dehaene (2015), 'Representation of Numerical and Sequential Patterns in Macaque and Human Brains', *Current Biology*, 25 (15): 1,966–74

Wang, X. J. (2001), 'Synaptic Reverberation Underlying Mnemonic Persistent Activity', *Trends in Neuroscience*, 24: 455–63

Wapner, W., T. Judd and H. Gardner (1978), 'Visual Agnosia in an Artist', *Cortex*, 14: 343–64

Ward, J. (2013), 'Synesthesia', *Annual Review of Psychology*, 64: 49–75

Warrier, V., V. Chee, P. Smith et al. (2015), 'A Comprehensive Meta-Analysis of Common Genetic Variants in Autism Spectrum Conditions', *Molecular Autism*, 6 (1): 49

Watson, J. B. (1930), *Behaviorism*, New York, NY: W. W. Norton/Chicago, IL: University of Chicago Press

Wechsler, D. (1944), *The Measurement of Adult Intelligence*, Baltimore, MD: Williams & Wilkins

Wegner, D. M. (2003), 'The Mind's Best Trick: How We Experience Conscious Will', *Trends in Cognitive Sciences*, 7 (2): 65–9

Wegner, D. M. and T. Wheatley (1999), 'Apparent Mental Causation: Sources of the Experience of Will', *American Psychologist*, 54 (7): 480–92

Wei, X., J. W. Yu, P. Shattuck et al. (2013) 'Science, Technology, Engineering, and Mathematics (STEM) Participation Among College Students with an Autism Spectrum Disorder', *Journal of Autism and Developmental Disorders*, 43 (7): 1,539–46

Weiskrantz, L. (1956), 'Behavioral Changes Associated with Ablation of the Amygdaloid Complex in Monkeys', *Journal of Comparative and Physiological Psychology*, 49: 381–91

Weisskopf, M. G., P. E. Castillo, R. A. Zalutsky and R. A. Nicoll (1994), 'Mediation of Hippocampal Long-Term Potentiation by Cyclic AMP', *Science*, 265: 1,878–82

Weissman, I. L. (2002), 'The Road Ended Up at Stem Cells', *Immunological Reviews*, 185: 159–74

Werling, D. M., N. N. Parikshak and D. H. Geschwind (2016), 'Gene Expression in Human Brain Implicates Sexually Dimorphic Pathways in Autism Spectrum Disorders', *Nature Communications*, 7: 10,717

Wernicke, C. (1874), *Der Aphasische Symptomenkomplex*, Breslau: Max Cohn & Weigert

Wertz, R. T., L. L. LaPointe and J. C. Rosenbek (1984), *Apraxia of Speech in Adults: The Disorder and its Management*, Orlando, FL: Grune and Stratton

West, M. J. (1990), 'Stereological Studies of the Hippocampus: A Comparison of the Hippocampal Subdivisions of Diverse Species Including Hedgehogs, Laboratory Rodents, Wild Mice and Men', *Progress in Brain Research*, 83: 13–36

Whalen, P. J., J. Kagan, R. G. Cook et al. (2004), Human Amygdala Responsivity to Masked Fearful Eye Whites', *Science*, 306: 2,061–6

Whalen, P. J. and E. A. Phelps (2009), *The Human Amygdala*, New York, NY: Guilford Press

Wheeler, M. E., S. E. Petersen and R. L. Buckner (2000), 'Memory's Echo: Vivid Remembering Reactivates Sensory-Specific Cortex', *Proceedings of the National Academy of Sciences of the United States of America*, 97: 11,125–9.

Wheelwright, S. J., B. Auyeung, C. Allison and S. Baron-Cohen (2010), 'Defining the Broader, Medium and Narrow Autism Phenotype Among Parents Using the Autism Spectrum Quotient (AQ)', *Molecular Autism*, 1: 10

Wheelwright, S. J., S. Baron-Cohen, N. Goldenfeld et al. (2006), 'Predicting Autism Spectrum Quotient (AQ) from the Systemizing Quotient-Revised (SQ-R) and Empathy Quotient (EQ)', *Brain Research*, 1,079 (1), 47–56

White, N. M. and R. J. McDonald (2002), 'Multiple Parallel Memory Systems in the Brain of the Rat', *Neurobiology of Learning and Memory*, 77: 125–84

White, T. D., B. Asfaw, Y. Beyene et al. (2009), 'Ardipithecus ramidus and the Paleobiology of Early Hominids', *Science*, 326 (5,949): 75–86

Whitlock, J. R., A. J. Heynen, M. G. Shuler and M. F. Bear (2006), 'Learning Induces Long-Term Potentiation in the Hippocampus', *Science*, 313: 1,093–7

Williams, M. A., A. P. Morris, F. McGlone et al. (2004), 'Amygdala Responses to Fearful and Happy Facial Expressions Under Conditions of Binocular Suppression', *Journal of Neuroscience*, 24 (12): 2,898–904

Williamson, L. (1978), 'Infanticide: An Anthropological Analysis', in M. Kohl (ed.), *Infanticide and the Value of Life* (pp. 61–75), Prometheus Books

Wilson, T. D., D. A. Reinhard, E. C. Westgate et al. (2014), 'Just Think: The Challenges of the Disengaged Mind', *Science*, 345 (6,192): 75–7

Windey, B., W. Gevers and W. Cleeremans (2013), 'Subjective Visibility Depends on Level of Processing', *Cognition*, 129 (2): 404–9

Wise, R. J. S., S. K. Scott, S. C. Blank et al. (2001), 'Separate Neural Subsystems Within "Wernicke's Area"', *Brain*, 124: 83–95

Wixted, J. T. and L R. Squire (2011), 'The Medial Temporal Lobe and the Attributes of Memory', *Trends in Cognitive Science*, 15: 210–17

Wolman, D. (2012), 'The Split Brain: A Tale of Two Halves', *Nature*, 483: 260–3

Wood, W. and D. Rünger (2016), 'Psychology of Habit', *Annual Review of Psychology*, 67: 289–314

Wrangham, R. W. (2010), *Catching Fire: How Cooking Made Us Human*, Profile

Wright, E. (ed.) (2008), *The Case for Qualia*, Cambridge, MA: MIT Press

Xu, W. and T. C. Sudhof (2013), 'A Neural Circuit for Memory Specificity and Generalization', *Science*, 339 (6,125): 1,290–5

Yang, J., S. Li, X. B. He et al. (2016), 'Induced Pluripotent Stem Cells in Alzheimer's Disease: Applications for Disease Modeling and Cell-Replacement Therapy', *Molecular Neurodegeneration*, 11 (1): 39

Yeatman, J. D., R. F. Dougherty, M. Ben-Shachar and B. A. Wandell (2012), 'Development of White Matter and Reading Skills', *Proceedings of the National Academy of Sciences of the United States of America*, 109 (44): E3045–53

Yoshida, K., Y. Go, I. Kushima et al. (2016), 'Single-Neuron and Genetic Correlates of Autistic Behavior in Macaque', *Sciences Advances*, 2 (9): e1600558

Young, L., A. Bechara, D. Tranel et al. (2010), 'Damage to Ventromedial Prefrontal Cortex Impairs Judgment of Harmful Intent', *Neuron*, 65: 845–51

Zaehle, T., T. Wüstenberg, M. Meyer and L. Jäncke (2004), 'Evidence for Rapid Auditory Perception as the Foundation of Speech Processing: A Sparse Temporal Sampling fMRI Study', *European Journal of Neuroscience*, 20 (9): 2,447–56

Zaidel, E. (1990), 'Language Functions in the Two Hemispheres Following Complete Commissurotomy and Hemispherectomy', in: F. Boiler and J. Grafman (eds), *Handbook of Neuropsychology*, New York, NY: Elsevier

Zaki, J., N. Bolger and K. Ochsner (2008), 'It Takes Two: The Interpersonal Nature of Empathic Accuracy', *Psychological Science*, 19 (4): 399–404

Zamzow, R. (2019), 'Rethinking Repetitive Behaviours in Autism', *Spectrum News*, www.spectrumnews.org/features/deep-dive/rethinking-repetitive-behaviors-in-autism/

Zeki, S. (2003), 'The Disunity of Consciousness', *Trends in Cognitive Sciences*, 7 (5): 214–18

—— (2007), 'A Theory of Micro-Consciousness', in M. Velmans and S. Schneider (eds), *The Blackwell Companion to Consciousness* (pp. 580–8), Oxford: Blackwell

Zeki, S., J. D. Watson, C. J. Lueck et al. (1991), 'A Direct Demonstration of Functional Specialization in Human Visual Cortex', *Journal of Neuroscience*, 11: 641–9

Zilles, K., N. Palomero-Gallagher and A. Schleicher (2004), 'Transmitter Receptors and Functional Anatomy of the Cerebral Cortex', *Journal of Anatomy*, 205 (6): 417–32

Zoghbi, H. Y. and M. F. Bear (2012), 'Synaptic Dysfunction in Neurodevelopmental Disorders Associated with Autism and Intellectual Disabilities', *Cold Spring Harbor Perspectives in Biology*, 4 (3)

Zurif, E. B., A. Caramazza and R. Meyerson (1972), 'Grammatical Judgments of Agrammatic Aphasics', *Neuropsychology*, 10: 405–17

Index